PROTEIN BIOCHEMISTRY, SYNTHESIS, STRUCTURE AND CELLULAR FUNCTIONS

# ZINC FINGERS

## STRUCTURE, PROPERTIES AND APPLICATIONS

# PROTEIN BIOCHEMISTRY, SYNTHESIS, STRUCTURE AND CELLULAR FUNCTIONS

Additional books in this series can be found on Nova's website under the Series tab.

Additional E-books in this series can be found on Nova's website under the E-book tab.

PROTEIN BIOCHEMISTRY, SYNTHESIS, STRUCTURE AND CELLULAR FUNCTIONS

# ZINC FINGERS

## STRUCTURE, PROPERTIES AND APPLICATIONS

ROLF CIOFANI
AND
LARS MAKRLÍK
EDITORS

Nova Science Publishers, Inc.
*New York*

For permission to use material from this book please contact us:
Telephone 631-231-7269; Fax 631-231-8175
Web Site: http://www.novapublishers.com



Additional color graphics may be available in the e-book version of this book.

**Library of Congress Cataloging-in-Publication Data**

Zinc fingers : structure, properties, and applications / editors, Rolf Ciofani, Lars Makrlmk.
p. ; cm. -- (Protein biochemistry, synthesis, structure, and cellular functions series)
Includes bibliographical references and index.
ISBN 978-1-62100-230-7 (hardcover)
I. Ciofani, Rolf. II. Makrlik, Lars. III. Series: Protein biochemistry, synthesis, structure, and cellular functions series.
[DNLM: 1. Zinc Fingers. 2. Protein Conformation. QU 55.9] LC classification not assigned
547'.05661--dc23 2011032488

*Published by Nova Science Publishers, Inc. † New York*

# Contents

# Preface

Zinc fingers are small protein structural motifs that can coordinate one or more zinc ions to help stabilize their folds. They can be classified into several different structural families and typically function as interaction modules that bind DNA, RNA, proteins or small molecules. In this book, the authors present current research in the study of the structure, properties and applications of zinc fingers. (Imprint: Nova Biomedical)

Chapter 1 - The CCCTC-binding factor (CTCF) is a highly conserved and ubiquitously expressed 11 zinc-finger DNA-binding protein among vertebrates, which is implicated in genome-wide organization of chromatin architecture and non-canonical gene regulatory functions. Direct evidence for CTCF-mediated higher-order chromatin structure has been reported at several developmentally regulated gene loci, including the mouse *β-globin* and *H19/Igf2* loci and the human *MHCII* locus. Cell-type-specific regulation of chromatin looping possibly arises through cooperation between the ubiquitous architectural factor CTCF and lineage-specific transcription factors.

Chapter 2 - Nucleocapsid (NC) proteins of retroviruses like HIV, derived from the proteolysis of the *gag*-encoded GAG polyprotein, are ≈ 50-amino acids long proteins that play pivotal roles in the viral lifecycle. These proteins contain one major functional domain consisting of two small zinc finger-like motifs named CCHC zinc knuckles conformed by the sequence Cys-$X_2$-Cys-$X_4$-His-$X_4$-Cys, where X can be any amino acid. The change of critical zinc binding amino acids; i.e., the three Cys or the His, to non-zinc binding residues results in virions that are defective in RNA packaging and replication. Interestingly, several proteins from eukaryotic cells contain from one to nine CCHC zinc knuckles and play essential functions in the metabolism of nucleic acids. Although NC proteins were widely studied, scant information exists about such cellular proteins. Here the authors classify cellular CCHC zinc knuckle-containing proteins into two main groups. One group consists of proteins containing one to three CCHC zinc knuckles interspersed among other functional motifs. The other group is conformed by proteins that contain four or more tandem repeated CCHC zinc knuckles with relative short linkers between them, forming a multi-zinc knuckle core domain. Moreover, the authors revise the function and manner of action of the most representative eukaryotic cellular CCHC-zinc knuckle proteins of each group, and carry out a structural and functional comparison between retroviral and cellular eukaryotic CCHC zinc knuckle-containing proteins. Finally, they point out some issues that should be explored to better understand the function of CCHC zinc knuckles in eukaryotic cellular proteins.

Chapter 3 - RNA is central to many cellular and viral functions. Agents that interact with specific RNAs offer means of intervention and serve as tools for functional studies. The authors have demonstrated the successful targeting of an HIV RNA substructure (RREIIB) by zinc finger proteins. These proteins were initially evolved by phage display, subsequently pruned to a single zinc finger and further enhanced by a structure and thermodynamic guided design. The structure of the zinc finger and the changes on the RNA target induced upon binding were investigated using an array of techniques including gel shift experiments, NMR spectroscopy and isothermal titration calorimetry. In addition to the α-helical portion, other secondary structures of the zinc finger also contribute to the specific binding. Insights from the thermodynamic analysis provide a detailed understanding of the molecular processes of the interaction and highlighted the crucial role of water in the binding process. Taking the effect of favorable interactions and water into account, zinc fingers with higher RREIIB-Tr binding affinity were generated. The authors' findings suggest strategies for enhancing the binding affinities of other biomolecular systems.

Chapter 4 - During evolution, the origin of new characteristics is accompanied by the appearance of novel genes. These genes, namely regulatory genes, can modulate effector genes located downstream in the functional pathways. Therefore, quantitative and qualitative modifications of these regulatory genes result in impressive changes in several biological processes. New biological functions also occur through the production of new protein variants. The modular structure of these proteins has allowed them to develop the capability to interact with DNA and RNA and with other protein partners, thus turning on or off regulatory pathways. A combination of ancient and modern modules can originate a complex protein repertoire bringing about the loss or gain of new specific functions in the various animal species.

Proteins able to play these regulatory roles, in most cases, belong to the transcription factor family.

Chapter 5 - Proteins containing zinc finger domains are highly abundant in eukaryotes. Of these, the CCCH class of motifs, where a zinc ion is postulated to be coordinated by three cysteine and one histidine residues, are present in a large number of proteins. Recent studies have implicated the members of this class in a wide variety of cellular functions ranging from developmental activities to amelioration of biotic and abiotic stresses. Animal CCCH zinc finger proteins recognize destabilizing AU-rich elements in the 3'-untranslated regions of messenger RNAs targeting them for degradation. Genetic analysis of plant proteins indicates their involvement in mRNA processing, floral morphogenesis, senescence, hormone-mediated environmental, biotic and abiotic stress responses. Furthermore, plant proteins exhibit differential expression patterns indicating diverse biological roles. Several plant proteins have similar functions and properties, such as trafficking between the nucleus and cytoplasm, and co-localizing with markers of processing bodies (PBs) or stress granules, although such proteins differ in zinc finger motif number they possess. Biochemical studies reveal that individual motifs of tandem fingers might participate in protein-protein interactions, nucleic acid binding, contributing ribonuclease activity or even calmodulin-mediated calcium signaling. This chapter summarizes the recent advances and current thinking of this protein family in plants including genome level characterization, classification, nucleic acid binding features, site directed mutagenesis based functional characterization, and finally, the future directions to fill the gaps in structural information.

Chapter 6 - The C2H2-type zinc finger motif is one of the most major DNA binding motifs in human transcription factors. The motif commonly contains the sequence of the type, (Tyr, Phe)-X-Cys-$X_{2-4}$-Cys-$X_3$-Phe-$X_5$-Leu-$X_2$-His-$X_{3-5}$-His, where X represents relatively non-conserved amino acids. Each finger folds into a compact $\beta\beta\alpha$ structure by coordination of a zinc ion with the conserved two cysteine (C) and two histidine(H) residues in a tetrahedral fashion. This fold is stabilized by hydrophobic interactions formed by the conserved hydrophobic amino acid residues. Though the Zn(II) ligands, two cysteine and two histidine residues, and the hydrophobic amino acid residues are highly conserved among the C2H2-type zinc fingers, their contribution to structure and functions is not clearly understood. We have created zinc finger mutants, in which the highly conserved amino acid residues are substituted to other amino acid residues. Mutation of the conserved hydrophobic amino acids on the DNA recognition helix or β-turn clarified the relationship between formation of a hydrophobic core and DNA binding function. On the other hand, ligand substituted HHHH-, CHHH-, and HCHH-type zinc fingers, in which cysteine residue(s) is mutated to histidine residue(s), showed hydrolytic ability and cleaved DNA in a sequence specific manner. They would be a new type of nucleases. In addition, the authors have successfully created the Zn(II)-dependent DNA binding domain, CDHH zinc finger, by reducing the Zn(II) binding affinity via ligand substitution of the second cysteine by aspartic acid. An artificial transcription factor with a CDHH-type zinc finger as a DNA binding domain showed Zn(II)-responsive transcriptional activity inside living cells. Here, they review the new functions of zinc fingers revealed by substitution of highly conserved residues.

Chapter 7 - Sp1 is a member of the zinc finger (ZF) family of transcription factor proteins which bind to GC-rich sequences, known as GC-boxes. Sp1 is a ubiquitous transcription factor and plays role in the regulation of expression of many genes. It has been shown *in vitro* that the promoter activity of human *SGLT1* gene is regulated by Sp1. This transcription factor also binds to a GC-box upstream of mouse *SGLT2* promoter. These genes encode proteins that are involved in transport of extracellular glucose into the cells. In the kidney, SGLT1 and SGLT2 are expressed in proximal tubule cells of the nephron and play crucial roles in reabsorption of glucose from glomerular filtrate. Chronic exposure of human to cadmium results in accumulation of this toxic metal in the kidney cortex which can cause down-regulation of glucose reabsorption and development of the symptom of glucosuria. The mechanism of cadmium-induced glucosuria may involves the inhibitory effect of this metal on the DNA binding activity of Sp1 that can mediate the transcriptional down-regulation of *SGLT1* and *SGLT2* gene expression.

In: Zinc Fingers
Editors: R. Ciofani and L. Makrlik
ISBN 978-1-62100-230-7


*Chapter 1*

# The Zinc-Finger DNA-Binding Protein CTCF: A Crucial Regulator of Gene Expression in T Lymphocytes

***Claudia Ribeiro de Almeida and Rudi W. Hendriks****
Department of Pulmonary Medicine,
Erasmus MC, 3000 CA Rotterdam,
The Netherlands

## Summary

The CCCTC-binding factor (CTCF) is a highly conserved and ubiquitously expressed 11 zinc-finger DNA-binding protein among vertebrates, which is implicated in genome-wide organization of chromatin architecture and non-canonical gene regulatory functions. Direct evidence for CTCF-mediated higher-order chromatin structure has been reported at several developmentally regulated gene loci, including the mouse *β-globin* and *H19/Igf2* loci and the human *MHCII* locus. Cell-type-specific regulation of

*E-mail address: r.hendriks@erasmusmc.nl, phone: ++31-10-7043700, fax: ++31-10-7044728

chromatin looping possibly arises through cooperation between the ubiquitous architectural factor CTCF and lineage-specific transcription factors.

In this chapter, we focus on the role of CTCF in the immune system, in particular the T lymphocyte lineage, both during development and in mature effector T cells.

Lymphocytes comprise the adaptive arm of the immune system and have the unique capacity to respond to a wide array of pathogens, due to their unique and virtually limitless capacity to specifically recognize antigens. This relies on the generation of a wide repertoire of antigen receptors during lymphocyte development.

Additionally, effective immune responses against pathogens depend on differentiation of naïve lymphocytes into mature effector cells, once they encounter their cognate antigen. Proper lymphocyte development as well as differentiation into functional effector subsets depends on the expression of lineage-associated genes, which is controlled by the combinatorial activity of multiple signaling pathways and transcription factors in a complex gene regulatory network.

The formation of a diverse repertoire of antigen receptors during lymphocyte development is a tightly regulated process. During this process, which is referred to as V(D)J recombination, recombination takes place between large genomic loci, containing over hundred V gene segments. Similarly, when naïve T cells become activated they undergo differentiation to give rise to various effector cell fates in a program that is dependent on coordinate regulation of complex cytokine gene loci.

These cytokines are crucial mediators of inflammation that when inappropriately regulated can initiate immune pathologies e.g. asthma or autoimmune diseases, including rheumatoid arthritis or multiple sclerosis. At these complex loci, regulation of gene expression involves the coordinate binding of lineage-specific transcription factors to *cis*-regulatory elements, such as promoters, enhancers, silencers or locus control regions, located over considerable genomic distances.

Emerging evidence suggests that long-range interactions between *cis*-regulatory elements, by looping of the intervening DNA fragment, are part of the mechanism of gene regulation at these developmentally regulated loci. Here, we describe recent findings that unravel the function of CTCF throughout T-lymphocyte development and effector T helper cell differentiation.

# General Introduction into the Biology of CTCF: A Multivalent Factor

CTCF is an ubiquitously expressed and highly conserved 11-zinc finger DNA-binding protein implicated in a variety of regulatory functions. These include transcriptional activation and repression, enhancer blocking and/or barrier gene insulation, hormone-responsive gene silencing, genomic imprinting, X chromosome inactivation and long-range chromatin interactions (Ohlsson et al., 2001; Phillips and Corces, 2009). Recent studies have mapped genome-wide CTCF occupancy and distribution patterns in multiple divergent cell types by employing chromatin immunoprecipitation (ChIP) in combination with tilling arrays (ChIP-chip) or high-throughput sequencing (ChIP-Seq). In human fibroblasts 13,804 CTCF-binding sites were reported (Kim et al., 2007) using ChIP-chip, while ChIP-Seq analysis identified 20,262 CTCF-binding sites in resting human $CD4^+$ T cells (Barski et al., 2007). ChIP-Seq analysis revealed 39,609 CTCF-binding sites in mouse embryonic stem cells (Chen et al., 2008), as well as 19,308 and 19,572 in HeLa and Jurkat cells, respectively (Cuddapah et al., 2009).

Importantly, these studies enabled the identification of a ~11-15 bp core consensus sequence for CTCF-binding sites, although initially CTCF was shown to bind to highly divergent 50-60 bp sequences through the combinatorial use of different sets of its 11 zinc fingers (Ohlsson et al., 2001). In fact, it was proposed that "during the formation of a CTCF-DNA complex both DNA and CTCF allosterically customize their conformation to engage different zinc fingers either for making base contacts or to make a target-specific surface that determines interactions between CTCF and other proteins" (Ohlsson et al., 2001). This property would be the basis of CTCF versatility or multiple functional roles in gene regulation. In line with this, CTCF binding partners that have been identified to date are involved in multiple processes in the nucleus and include transcription factors (Yy1, Kaiso and YB-1) (Chernukhin et al., 2000; Defossez et al., 2005; Donohoe et al., 2007), chromatin modifying proteins (Sin3a, CHD8, Suz12) (Ishihara et al., 2006; Li et al., 2008; Lutz et al., 2000), the nucleolar protein nucleophosmin (Yusufzai et al., 2004), RNA polymerase II (Chernukhin et al., 2007) and poly(ADP-ribose) polymerase-1 (PARP1) (Yu et al., 2004).

Genome-wide CTCF binding patterns suggest a role for CTCF distinct from that of canonical transcription factors (Phillips and Corces, 2009). CTCF distribution strongly correlates with gene density but not with transcriptional start sites and the binding pattern of CTCF does not appear to predict cell-specific gene expression levels (Chen et al., 2008). Domains with few or no CTCF-binding sites tend to include clusters of transcriptionally co-regulated genes or developmentally related genes (for example, olfactory receptor gene clusters or keratin-associated protein gene clusters), whereby these regions are often flanked by CTCF-binding sites (Kim et al., 2007). CTCF-rich regions contain genes displaying extensive alternative promoter usage (for example, *Tcrβ*, *Tcrα/δ* and *Igλ* L chain loci) (Kim et al., 2007).

Prior to genome-wide queries, many studies have suggested that CTCF is the archetypal vertebrate protein that binds insulator sequences. Insulators are genomic elements that affect gene expression by blocking the inappropriate communication between neighboring regulatory elements (enhancer-blocking insulators) and preventing the spread of repressive heterochromatin from adjacent sequences (barrier insulators) (Wallace and Felsenfeld, 2007). The first link between CTCF and enhancer-blocking gene insulation was proposed on the basis of the discovery that CTCF binds to the 5'HS4 insulator sequence upstream of the chicken *β-globin* locus (Bell et al., 1999) and to another insulator sequence at the 3'end of this domain (Saitoh et al., 2000). CTCF mediated insulator activity has been predicted at several other sites including the *H19/Igf2* genes imprinted control region (ICR) (Bell and Felsenfeld, 2000; Hark et al., 2000; Kanduri et al., 2000; Szabo et al., 2000a), the *DM1* locus (Filippova et al., 2001) and the boundaries of domains that escape X chromosome inactivation (Filippova et al., 2005). More recently, computational analyses have been employed to show that CTCF binds to highly conserved non-coding elements across 12 mammalian species, which provided evidence supporting a global role for CTCF as an insulator protein (Xie et al., 2007). Although CTCF is generally considered to function via enhancer-blocking mechanisms, genome-wide data sets also uncovered a small proportion of CTCF-binding sites at the boundaries between active and repressive chromatin domains, which can reflect a role for CTCF in barrier insulation (Barski et al., 2007; Cuddapah et al., 2009).

## CTCF-Mediated Chromatin Loops: Gene Insulation and Beyond

Enhancer-blocking CTCF-driven insulation may occur through the formation of long-range chromatin interactions or chromatin loops. A recent study has provided the first direct evidence that loop formation can occur via contacts between two CTCF-bound insulators *in vivo* (Hou et al., 2008). This was shown, by inserting the human *β-globin* HS5 (the orthologue of the CTCF-dependent chicken HS4 insulator) between the mouse *β-globin* LCR and downstream genes, which results in the formation of a new chromatin loop with endogenous HS5 that topologically isolates the LCR and disrupt *β-globin* gene transcription.

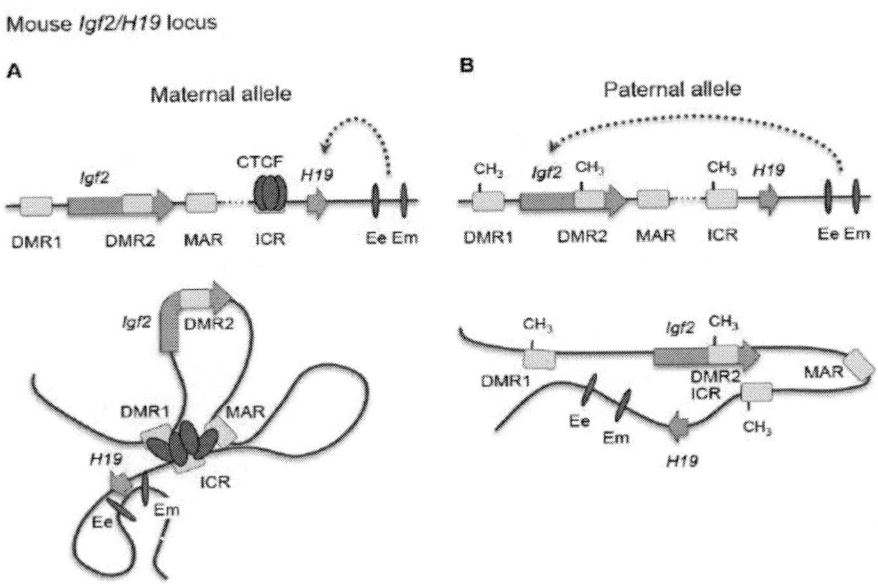


Figure 1. Schematic representation of the mouse *Igf2/H19* locus. The maternally expressed noncoding *H19* gene (A) is located approximately 90 Kbp downstream from the gene encoding Insulin-like growth factor 2 (*Igf2*) that is expressed exclusively from the paternal allele (B). The imprinting control region (ICR) ~2 Kbp upstream of *H19* contains four CTCF binding sites and is essential for the regulation of the entire locus. Differentially methylated regions (DMRs) - DMR1 upstream of *Igf2* promoters and DMR2 within *Igf2* exon 6 - act in concert to regulate reciprocal, allele-specific expression patterns from a shared set of enhancers downstream of the H19 gene at 8 Kbp (Ee, endodermal tissue enhancer) and 25 Kbp (Em, mesodermal tissue enhancer). Allele-specific patterns of CTCF binding, DNA methylation (–CH3) and chromatin looping account for parent-of-origin *H19* and *Igf2* gene expression patterns during development. Figure adapted from Phillips and Corces, 2009.

Moreover, at the imprinted *H19/Igf2* locus CTCF-mediated enhancer-blocking insulation have been demonstrated to occur via long-range chromatin interactions in the native genomic context (Kurukuti et al., 2006; Murrell et al., 2004) (Figure 1). CTCF is considered to play a critical role in the establishment and maintenance of allele-specific DNA methylation imprints and parent-of-origin *H19* and *Igf2* gene expression patterns during development. Four methylation-sensitive CTCF-binding sites are located in the ICR, immediately upstream of the *H19* gene.

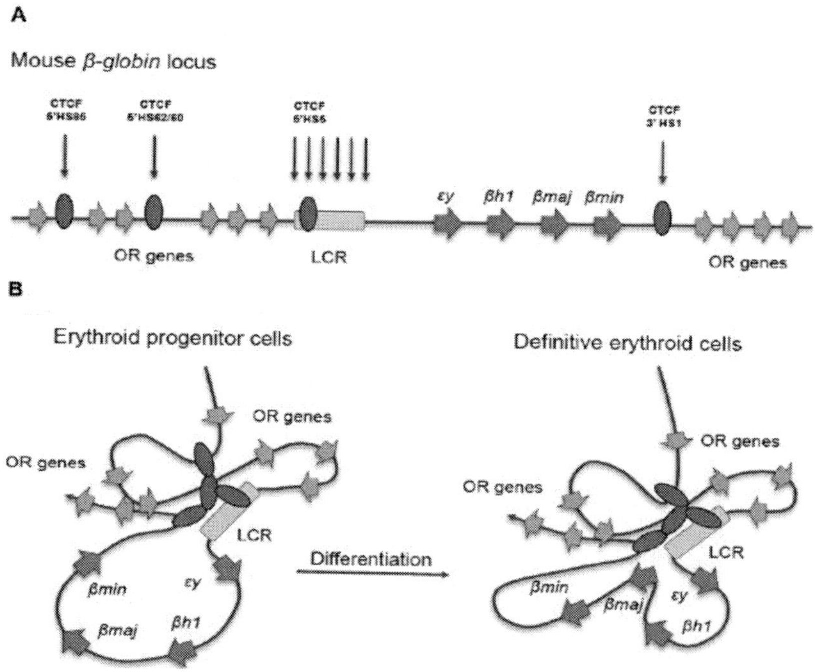


Figure 2. Schematic representation of the mouse *β-globin* locus. (A) Four globin genes are embedded within a larger olfactory receptor (OR) gene cluster. Developmentally regulated globin expression (*εy* and *βh1* in primitive erythroid cells; *β-major* (*βmaj*) and *β-minor* (*βmin*) in definitive erythroid cells) is regulated in part by a series of *cis*-regulatory elements surrounding the locus (indicated by arrows). An upstream locus control region (LCR) containing six DNaseI hypersensitive sites (HS) is required for high-level transcription. Three CTCF binding sites have been identified upstream (5'HS85, 5'HS62/60 and 5'HS5) and one 20 Kbp downstream (3'HS1) of *globin* genes. (B) Cell type-specific CTCF-mediated intrachromosomal interactions in the mouse *β- globin* locus. Although loops are illustrated by CTCF multimerization, it is not clear whether CTCF binds directly to each site or through interaction with other proteins that are not represented for simplicity. In definitive erythroid cells full expression of, for example, the *βmaj* gene depends on stable LCR-gene contacts. Figure adapted from Phillips and Corces, 2009.

On the maternal allele CTCF binds to the unmethylated ICR to form a tightly coiled chromatin loop structure around the *Igf2* gene, thereby making the *Igf2* promoters inaccessible to the enhancers downstream of *H19*. By contrast, on the paternal allele the ICR is methylated, which abrogates CTCF binding and ICR-mediated insulation resulting in functional communication between the *Igf2* promoters and the enhancers in order to activate *Igf2* expression. *H19* is only expressed from the maternal allele, since loss of CTCF binding to the methylated ICR is thought to silence *H19* on the paternal allele due to promoter methylation.

CTCF-based intrachromosomal interactions between distal regulatory elements have also been reported at other developmentally regulated gene clusters, such as the *β-globin* and the *MCHII* loci. CTCF binds to three sites upstream (5'HS85, 5'HS62/60 and 5'HS5 in the LCR) and one downstream (3'HS1) of the mouse *β-globin* locus in a cell-type specific manner (Bulger et al., 2003; Farrell et al., 2002; Splinter et al., 2006) (Figure 2).

During erythroid differentiation, CTCF-bound regulatory sequences throughout the *β-globin* locus came into close spatial proximity to form an "active chromatin hub" where stage-specific *globin* genes also interact with the LCR resulting in looping-out of transcriptionally silent isoforms (Palstra et al., 2003; Splinter et al., 2006; Tolhuis et al., 2002).

Conditional CTCF deletion or mutation of the 3'HS1 CTCF-binding site resulted in disruption of CTCF contacts in erythroid progenitors but surprisingly had no effects on the kinetics of levels of *globin* gene expression during erythroid differentiation (Splinter et al., 2006). One explantion could be the redundancy of chromatin contacts in the *β-globin* locus that may not be readily abrogated by mutations in one regulatory element. The functional significance of CTCF-mediated insulation at the 3'HS1 and 5'HS5 of the *β-globin* locus is also unclear. Genetic disruption of CTCF-binding to HS1 has no effect on the transcription of the neighboring olfactory receptor genes (Splinter et al., 2006) and deletion of 3'HS1 and 5'HS5 results in normal *globin* gene expression (Bender et al., 2006; Bender et al., 1998).

Likewise, at the human major histocompatibility complex (*MHCII)* locus CTCF may not function as a canonical enhancer-blocking insulator either. CTCF binds to the intergenic enhancer XL9 that co-regulates the expression of the MHCII genes *HLA-DRB1* and *HLA-DQA1* and shows enhancer-blocking activity *in vitro* (Majumder et al., 2008) (Figure 3).

In non-MHCII-expressing epithelial cells, stimulation with IFNγ results in the expression of the transcriptional co-activator CIITA and the concurrent formation of chromatin interactions between XL9 and the two divergent promoters upstream of *HLA-DRB1* and *HLA-DQA1*. Chromatin loop formation is required for gene expression and depends on XL9-bound CTCF, the non-DNA-binding co-activator CIITA and the transcription factor RFX constitutively bound to regulatory sequences within the proximal promoter of *HLA-DRB1* and *HLA-DQA1* genes. Therefore, in the human *MHCII* locus CTCF is involved in the control of gene expression through the ability to form transcriptionally functional chromatin loops. Nevertheless, we cannot rule out the possibility that XL9-bound CTCF is blocking inappropriate regulatory elements contained outside the *HLA-DRB1* and *HLA-DQA1* region.

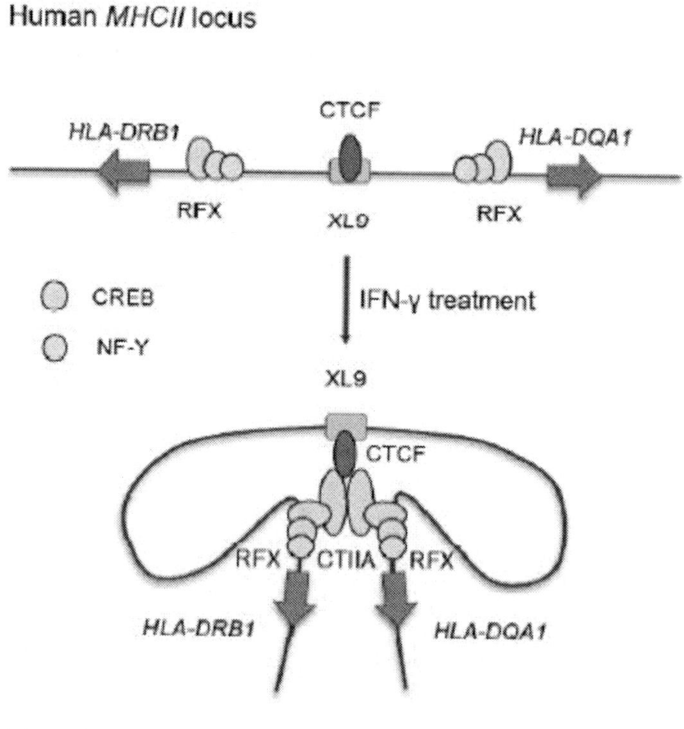


Figure 3. Schematic representation of the human *MHCII* locus. In the human *major histocompatibility complex class II* (*MHCII*) locus, CTCF binds to the XL9 enhancer element between two coregulated genes, *HLA - DRB1* and *HLA - DQA1*, driven by divergent promoters. Transcription factors RFX, CREB and NF-Y bind to regulatory sequences within the proximal promoter of *MHCII* genes when they are transcriptionally inactive. Interferon-γ (IFN-γ) treatment induces transcription in non-expressing cell types by upregulating the non-DNA binding coactivator CIITA, which subsequently forms an heterodimer with RFXCREB-NF-Y-bound promoters and the CTCF-bound enhancer in parallel with *HLADRB1* and *HLA - DQA1* gene activation. Figure adapted from Phillips and Corces, 2009.

# CTCF as a Genome-Wide Organizer of Chromatin Architecture

Mechanistic insight into CTCF-mediated chromatin loop formation has recently been uncovered by genome-wide studies showing that cohesin proteins are highly enriched at most CTCF-binding sites (Parelho et al., 2008; Rubio et al., 2008; Stedman et al., 2008; Wendt et al., 2008). CTCF is required for cohesin recruitment to these sites because depletion of CTCF leads to absence of cohesin binding. Because cohesin proteins are though to coalesce into a ring-like structure that mediates sister chromatid cohesion during mitosis, it is tempting to speculate that a similar mechanism could facilitate the stabilization of CTCF-based chromatin loops.

In support of this notion, cohesin proteins are required for CTCF-mediated long-range intrachromosomal interactions at the developmentally regulated *IFNg* locus (Hadjur et al., 2009). Functional interaction between cohesins and CTCF is also observed at sites where CTCF was previously shown to have enhancer-blocking insulator activity, such as the HS4 of the chicken *β-globin* locus and *H19/Igf2* locus ICR (Parelho et al., 2008; Wendt et al., 2008). The discovery that a high proportion of CTCF-binding sites are shared with cohesin provide a strong indication that cohesin and CTCF might act in concert at many genomic locations to facilitate the formation of long-range chromatin interactions.

Together with unique CTCF distribution patterns genome-wide these results implicate CTCF as a genome-wide organizer of chromatin architecture, with insulation and/or downstream effects in transcription a secondary consequence of the genomic context of the endogenous locus (Phillips and Corces, 2009). In line with this view, recent studies have implicated CTCF in interchromosomal interactions in the maternal (CTCF-bound) *H19/Igf2* ICR allele (Ling et al., 2006; Sandhu et al., 2009; Zhao et al., 2006) or between homologous X chromosomes during the process of X chromosome inactivation (Tsai et al., 2008; Xu et al., 2007).

Although the functional significance of CTCF-mediated inter-chromosomal contacts remains to be fully elucidated, it has been suggested that, at least at the *H19/Igf2* ICR, these may function to control epigenetic information *in trans* (Sandhu et al., 2009; Zhao et al., 2006). CTCF has also been suggested to demarcate the boundaries of "lamina-associated domains"

(LADs), genomic regions localized to the nuclear periphery via interaction with the nuclear lamina protein LaminB1 that correlate with markedly decreased gene expression (Guelen et al., 2008).

One example is the CTCF-binding site located ~10 kb upstream of the human *c-myc* gene that maps to a LAD border, which has been implicated in nuclear organization and locus protection from DNA methylation (Gombert et al., 2003; Gombert and Krumm, 2009). Genome-wide identification of CTCF-binding sites also revealed that a high proportion of CTCF-binding sites are present in exons, introns and 5' or 3' untranslated regions (UTRs), suggesting that intragenic and/or single gene loops may also have functional roles (Barski et al., 2007; Chen et al., 2008; Cuddapah et al., 2009; Kim et al., 2007; Xie et al., 2007). For example, in the human *c-Myc* gene one CTCF-binding site, found immediately downstream of the major P2 promoter, maps precisely within the region of RNA polymerase II pausing and release (Filippova et al., 1996). It is also plausible that intragenic and/or single gene loop formation facilitates the coordination of RNA processing or transcriptional re-initiation.

Nevertheless, the possibility remains that CTCF can function in the regulation of gene expression via looping-independent mechanisms by simply recruiting proteins involved in transcriptional activation and/or repression, as for example histone modifying proteins (Li et al., 2008; Lutz et al., 2000), chromatin remodeling factors (Ishihara et al., 2006) or RNA polymerase II (Chernukhin et al., 2007).

## CTCF: A Crucial Regulator of Gene Expression in Lymphocytes

Since lymphocyte development and effector function is crucially dependent on correct gene regulation of various large and complex gene loci, such as the immunoglobulin (Ig), T cell receptor (TCR), MHC and cytokine loci, the lymphocyte cell lineage comprises an ideal model system to unravel the molecular mechanisms by which CTCF regulates gene expression.

B- and T-lymphocytes are the central players of the adaptive immune system, which is characterized by gradual, though highly specific and effective immune responses against pathogens. Their unique and virtually limitless capacity to specifically recognize antigens relies on the generation of a wide repertoire of antigen receptors during lymphocyte development. Additionally, effective immune responses against pathogens depend on differentiation of naïve lymphocytes into mature effector cells, once they encounter their cognate antigen.

Proper lymphocyte development and differentiation into functional effector subsets depends on the expression of lineage-associated genes, which is controlled by the combinatorial activity of multiple signaling pathways and transcription factors in a complex gene regulatory network. In many developmentally regulated gene clusters, as for example the antigen receptor and T helper cytokine gene loci, gene regulation is particularly complex (Jhunjhunwala et al., 2009; Lee et al., 2006). This involves the coordinate binding of lineage-specific transcription factors to *cis*-regulatory elements, such as promoters, enhancers, silencers or locus control regions (LCRs), located over considerable genomic distances. Emerging evidence suggests that long-range interactions between *cis*-regulatory elements, by looping of the intervening DNA fragment, are part of the mechanism of gene regulation at these developmentally regulated loci.

Cell-type-specific regulation of higher-order chromatin structures possibly arises through cooperation between lineage-specific transcription factors and ubiquitously expressed proteins. For instance, the erythroid transcription factor Gata1 has been shown to be part of a complex composed of the hematopoietic factors Lmo-2, Tal1 and E2A and the ubiquitous nuclear protein Ldb1 (Meier et al., 2006; Rodriguez et al., 2005; Wadman et al., 1997), involved in long-range chromatin interactions at the mouse *β-globin* locus (Soler et al., 2010; Song et al., 2007). The ubiquitous architectural factor CTCF is a leading candidate for the regulation of gene expression at complex loci during lymphocyte development and differentiation into functional effector subsets.

## T Cell Development in the Thymus

T cells are unique among blood cells, as they require a dedicated organ for their development, the thymus. Hematopoietic progenitors must therefore leave the bone marrow (BM), traffic through the circulation and settle the thymus before developing into the T cell lineage.

Early T cell precursors in the thymus (lacking the cell surface markers CD4 and CD8 and therefore termed double-negative (DN) cells) are classically subdivided into four distinct developmental stages (DN1–DN4) by differential expression of the cell surface markers CD25 and CD44 (Godfrey et al., 1993) (Figure 4).

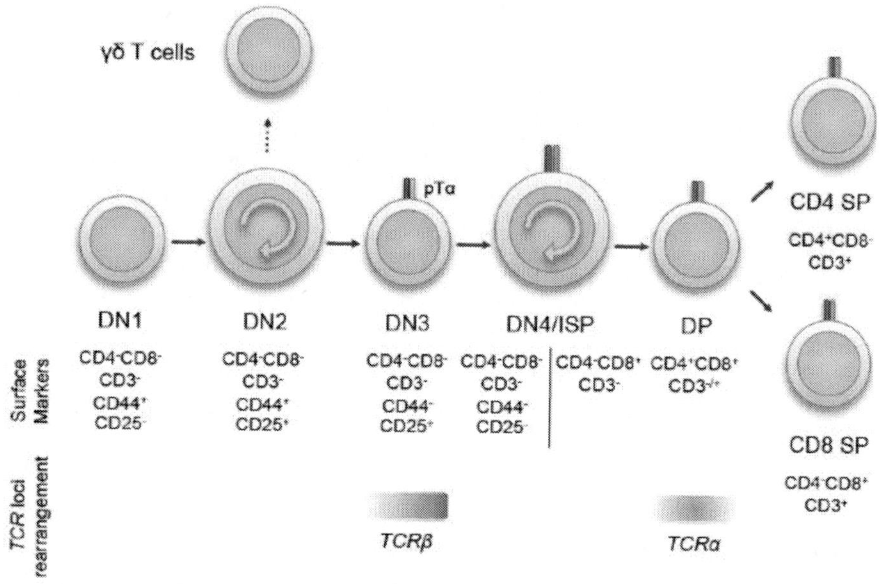


Figure 4. Schematic overview of T-cell developmental stages. T-cells develop in the thymus towards αβ or γδ T-cell lineages. Eight distinct αβ thymocyte developmental stages are identified on the basis of cell surface markers expression and the rearrangement status of *TCRβ* and *TCRα* loci. DN, double-negative; ISP, immature single positive; DP, double-positive; SP, single positive; pTα, pre-TCRα chain.

Further examination of the DN1 population revealed that efficient T lineage progenitor activity resides in a c-Kit$^{+}$CCR9$^{+}$ subset termed early thymic progenitors (ETPs) (Benz et al., 2008), which have very limited B cell potential and still retains the capacity to diverge into natural killer (NK), dendritic cell (DC) and myeloid lineages. Although the precise origin of ETPs remains controversial, it is likely that ETPs arise from the lymphoid-primed multipotent progenitors (LMPP) compartment in the BM (Chi et al., 2009).

The processes of T cell lineage specification and commitment are dependent on the coordinate actions of signaling cascades and transcription factors networks. However, the establishment of a T cell lineage program is remarkable for its dependence on a balance among multiple factors, none of which appears to act as a T lineage "master regulator" (Rothenberg et al., 2008; Rothenberg and Taghon, 2005).

Development through the "pro-T cell" stages – that is from the ETP to DN3 stage – is characterized by extensive proliferation, initiation of the T cell developmental program and restriction of alternative cell lineages. Coordinated migrations of pro-T cells through distinct thymic microenvironments (Petrie and Zuniga-Pflucker, 2007), provides changes in the combinations of receptor ligands (for example, Delta-like and Jagged family of Notch ligands and the c-Kit ligand stem cell factors (SCF)) and growth factors (such as IL-7) expressed by the thymic epithelium, which drive pro-T cell proliferation, survival and differentiation. Both c-Kit and IL-7R sustain most pro-T cell proliferation and survival (Balciunaite et al., 2005; Massa et al., 2006; Peschon et al., 1994; von Freeden-Jeffry et al., 1995).

Notch signaling is essential for several distinct aspects of early T cell development (Yuan et al.), including the onset of the T cell lineage program (Pui et al., 1999; Radtke et al., 1999), suppression of B, NK, DC and myeloid lineages potentials (Bell and Bhandoola, 2008; Ciofani and Zuniga-Pflucker, 2005; Pui et al., 1999; Schmitt et al., 2004; Wada et al., 2008), survival, proliferation and/or differentiation of DN3 cells (Ciofani and Zuniga-Pflucker, 2005).

Although Notch signals are critical to promote T cell fate, they most likely act in concert with a core group of transcription factors in the T cell specification and commitment process including: GATA-binding protein 3 (Gata3); avian myeloblastosis virus oncogen homolog (Myb); runt-related transcription factor 1-core binding factor β (Runx1-CBFβ) complexes; bHLH E-proteins E2A and HeLa E-box binding factor (HEB); T-cell factor 1 (Tcf1); growth factor independent 1 (Gfi1) and Ikaros (Rothenberg et al., 2008; Rothenberg and Taghon, 2005). Other transcription factors which expression is more dynamically regulated during T cell developmental stages collaborate with the core group of transcription factors to induce stage-specific gene expression programs. These include: 1) pre-thymically inherited non-T cell transcription factors (PU.1, stem cell leukemia (SCL)/Tal1, Gata2 and CCAAT-enhancer binding protein α (C/EBPα)) (Tydell et al., 2007) that most likely sustain the lineage plasticity in early stages of T cell development and which expression is downregulated before the DN3 stage, when T cell precursors are committed to the T cell lineage and can no longer develop into alternative cell types; 2) transcription factors expressed at the ETP to DN2 transition (Bcl11b, HEBalt (an alternative promoter isoform of HEB) and glioma-associated oncogene 2 (Gli2)) (Tabrizifard et al., 2004; Tydell et al., 2007), when most T cell identity genes (encoding for the precursor-TCR (pre-TCR) assembly complex - CD3ε, CD3γ and pre-TCRα chain (pTα), signaling components - kinases, phosphatases and adaptor proteins such as LCK, ζ-chain associated protein kinase of 70 kDa (Zap70) and linker of activated T cells (Lat), RAG proteins and IL-7Rα) begin to be upregulated (Tabrizifard et al., 2004; Taghon et al., 2005; Tydell et al., 2007); 3) and transcription factors expressed at the DN2 to DN3 transition (Ets1, Ets2, lymphoid-enhancer-binding factor 1 (Lef1), SpiB and Hes1) (Tabrizifard et al., 2004; Taghon et al., 2006; Taghon et al., 2005; Tydell et al., 2007), when most T cell identity genes are fully activated and commitment to the T cell lineage is completely and stably established.

Further differentiation of T cells depends on successful assembly of the T-cell receptor (TCR). Each receptor consists of a TCRα and a TCRβ chain, encoded by *Tcrα* and *Tcrβ* loci, respectively (Figure 4). Most mature T cells express a TCRαβ on their cell surface (αβ T-cells) and these are the T cells involved in the classic adaptive immune response. However, a distinct and minor subset of T cells (γδ T-cells) with largely elusive functional role in the immune system bears an alternative TCR made up of TCRγ and TCRδ chains, encoded by the *Tcrγ* and *Tcrδ* loci, respectively.

A large variety of TCR antigen specificities is achieved by a tightly regulated process of gene rearrangement initiated by RAG proteins at the *Tcr* loci (Jung et al., 2006; Krangel, 2009; Schlissel, 2004). During this process, which is referred to as V(D)J recombination, recombination takes place between multiple V (variable), D (diversity, only for *Tcrδ* and *Tcrβ* loci) and J (joining) gene segments. Selection of a V, (D) and J gene segment during gene rearrangement occurs at random and the large number of possible different combinations accounts for most of the diversity of the TCR repertoire. During T cell development, sequential rearrangement of *Tcrδ*, *Tcrγ* or *Tcrβ* and *Tcrα* loci is essential for the production of a functional TCR. During the DN2 to DN3 transition, rearrangements at the *Tcrδ, Tcrγ* and *Tcrβ* loci are initiated (Miosge and Zamoyska, 2007). Cells that productively rearrange the *Tcrδ* and *Tcrγ* loci may express a functional TCRγδ on the membrane, undergo a small proliferative burst and develop into γδ T-cells. Upon productive rearrangement of the *Tcrβ* locus, the TCRβ chain associates with the invariant pTα chain and CD3 molecules on the cell surface and forms the pre-TCR complex, which qualifies DN3 cells to enter the αβ T cell lineage.

Despite the presence of multiple CTCF-binding sites near the *Tcrα/δ* and *Tcrβ* loci (Garrett et al., 2005; Kim et al., 2007), we found that these loci undergo functional V(D)J recombination upon deletion of CTCF in thymocytes from *Lck-Cre Ctcf$^{f/f}$* mice (Heath et al., 2008), suggesting that CTCF is not absolutely essential for V(D)J recombination. Recent studies suggest that stronger signals delivered by the TCRγδ and relatively weaker pre-TCR signals determine the αβ/γδ-lineage choice (Haks et al., 2005; Hayes et al., 2005; Kreslavsky et al., 2008).

Pre-TCR signaling tests for the functionality of the TCRβ chain, a process called β-selection (von Boehmer, 2005). Cells that have passed this checkpoint are termed β-selected cells (or late DN3 cells, to distinguish them from early DN3 cells that are not yet selected) (Hoffman et al., 1996). The pre-TCR complex signals for *Tcrβ* locus allelic exclusion and DN3 cells survival, proliferation and differentiation. In particular, pre-TCR signals result in the downregulation of CD25 expression and progression of DN3 cells into the DN4 stage (Miosge and Zamoyska, 2007) (Figure 4).

Cells subsequently acquire both CD4 and CD8 co-receptors to become double-positive (DP) cells, with CD8 usually being expressed first in most mouse strains generating immature SP (ISP) cells. Therefore, the late DN3, DN4 and ISP stages consist of large cycling "pre-T cells". At the DP cell stage, proliferation is ceased before rearrangement at the *Tcrα* locus is initiated, which leads to deletion of the *Tcrδ* locus located in between $V_\alpha$ and $J_\alpha$ gene segments. If *Tcrα* gene recombination is productive, TCRαβ is expressed on the cell surface of DP cells. TCRαβ-bearing DP cells are tested for self-MHC recognition during the process of positive selection (He and Kappes, 2006). TCRαβ receptors recognizing MHC class I or class II molecules will develop into CD8 single positive (SP) or CD4 SP T cells, respectively (Singer et al., 2008). Two models have been elaborated to explain CD4/CD8 lineage commitment.

In the stochastic/selective model, lineage commitment is determined randomly, followed by a selection step that eliminates thymocytes that express the inappropriate thymocytes. The instructive/kinetic model proposes that qualitatively distinct signals initiated upon TCR engagement by MHC class I or class II ligands and duration of TCR signaling, instruct CD8 or CD4 lineage choice, respectively. DP thymocytes that fail to recognize MHC molecules die "by neglect", whereas potential self-reactive T lymphocytes are eliminated by apoptosis, a process called negative selection. When combined, these selection processes result in the generation of CD4 and CD8 SP thymocytes with TCRαβ receptors that can recognize non-self-antigens presented by MHC class II and I proteins, respectively. Mature SP cells exit the thymus and migrate via the bloodstream to peripheral lymphoid organs – such as the spleen, lymph nodes and mucosa associated lymphoid tissue (MALT; for example, Peyer's Patches) - as naïve recirculating $CD4^+$ T helper and $CD8^+$ cytotoxic T cells.

## Role for CTCF in T Cell Development in the Thymus

Our in vivo studies in the T cell lineage (Heath et al., 2008) have shown that CTCF is essential for the efficient proliferation of β-selected cells, in particular for their maturation from ISP to DP cells, as well as for TCR upregulation at the cell surface of DP thymocytes. To investigate its function *in vivo*, we generated mice with a conditional *Ctcf* knockout allele (Heath et al., 2008). In crosses with mice expressing Cre recombinase ubiquitously, we found that ablation of the *Ctcf* gene results in early embryonic lethality: no homozygous knock-outs were born from heterozygous crosses. Surprisingly, the ratio of wild-type to *Ctcf* heterozygous littermates was also higher than expected on a Mendalian basis in crosses among *Ctcf* heterozygous mice or between wild-type and *Ctcf* heterozygous mice. Therefore, these data suggest that CTCF is required in a dose-dependent manner. Tissue-specific inactivation of CTCF in thymocytes from *Lck-Cre* $Ctcf^{f/f}$ mice specifically hampered the differentiation of αβ T cells, particularly the proliferation of β-selected cells (DN4 and ISP cells) and their differentiation from ISP to DP cells. This differentiation block was associated with an accumulation of late DN and immature SP cells in the thymus of mice. These cells are normally large and actively cycling, and contain elevated amounts of CTCF. In *Ctcf* knockout animals, however, these cells are small and blocked in the cell cycle due to increased expression of the cyclin-CDK inhibitors p21 and p27. Taken together, our results show that CTCF is required in a dose-dependent manner and is involved in cell cycle progression of TCR αβ T cells in the thymus. We proposed that CTCF positively regulates cell growth in rapidly dividing thymocytes so that appropriate number of cells is generated before positive and negative selection in the thymus (Heath et al., 2008).

Observed cell type-specific or differentiation stage-specific changes in gene expression may explain the proliferation and differentiation defects observed in thymocytes deleted for CTCF. The β-selected cells from *Lck-Cre* $Ctcf^{f/f}$ mice have highly increased expression of the cyclin-CDK inhibitor encoding genes *p21* and *p27*. In contrast, in CTCF-deficient mouse embryonic fibroblasts expression of *p21*, but not of *p27*, is increased. CTCF binds to two adjacent sites in the first intron of *p21*, but no correlation between CTCF binding and *p21* expression is observed in T-cells (Heath et al., 2008). Therefore, whether CTCF is a direct transcriptional regulator of *p21* and/or *p27* remains to be seen. It is possible that CTCF cooperates with distinct transcription factors for positive or negative regulation of *p21* and/or *p27* expression in different cell-types. Similarly, expression of genes required for cell differentiation might depend on the cooperation between CTCF and lineage-specific transcription factors.

In line with this, ISP and DP cells from *Lck-Cre Ctcf$^{f/f}$* mice express low levels of TCRβ (Heath et al., 2008) and CTCF binding at the *Tcrβ* and *Tcrα/δ* loci has been previously reported (Gomos-Klein et al., 2007; Kim et al., 2007; Magdinier et al., 2004). The T-cell transcription factor Gata3 binds to *Tcr* enhancer elements and plays an important role in the regulation of T-ccll-specific expression of the *Tcrβ* and *Tcrα/δ* loci (Ho et al., 1991; Marine and Winoto, 1991). In this context, γδ T-cells are unique in the sense that CTCF is not required for their development or *in vitro* proliferation (Heath et al., 2008), which makes this cell population a prime candidate for genome-wide gene expression analysis.

## Differentiation of Effector T Cell Subsets

Mature CD4$^+$ or CD8$^+$ T cells that have not yet encountered their specific antigen recirculate between blood and peripheral lymphoid organs as naïve T cells. At sites of infection DCs take up antigen and are activated as part of the innate immune response, through receptors that signal the presence of pathogen constituents or by cytokines produced during the inflammatory response (Lambrecht and Hammad, 2010).

This induces their migration to local lymphoid tissue and their maturation into cells that are highly effective at presenting antigen to naïve T cells in the form of an antigen:MHC complex. Peptide antigens from intracellular pathogens that multiply in the cytoplasm are carried to the cell surface of DCs by MHC class I (MHCI) molecules and presented to CD8$^+$ T cells. Peptide antigens from pathogens multiplying in intracellular vesicles and those derived from extra-cellular pathogens, are carried to the cell surface by MHC class II (MHCII) molecules and presented to CD4$^+$ T cells. Mature DCs are distinguished by the expression of high levels of surface co-stimulatory molecules, such as CD80 and CD86 that bind to the CD28 receptor on T cells.

Simultaneous delivery of two independent signals by the same APC - an antigen-specific TCR signal and a co-stimulatory signal – is absolutely required for the activation of naïve T cells. This means that only activated APCs can initiate T cell responses, which is important because not all potentially self-reactive T cells are deleted in the thymus.

Naïve T cells recognizing self-antigens in the absence of co-stimulation are not activated and instead are induced to enter a state of non-responsiveness or anergy. T cell activation is characterized by clonal expansion of antigen-specific T cells and differentiation of $CD4^+$ or $CD8^+$ T cells into functionally distinct T helper or cytotoxic effector subsets, respectively. It is crucially dependent on the synthesis of IL-2 along with the α chain of the IL-2 receptor (also known as CD25) by activated T cells.

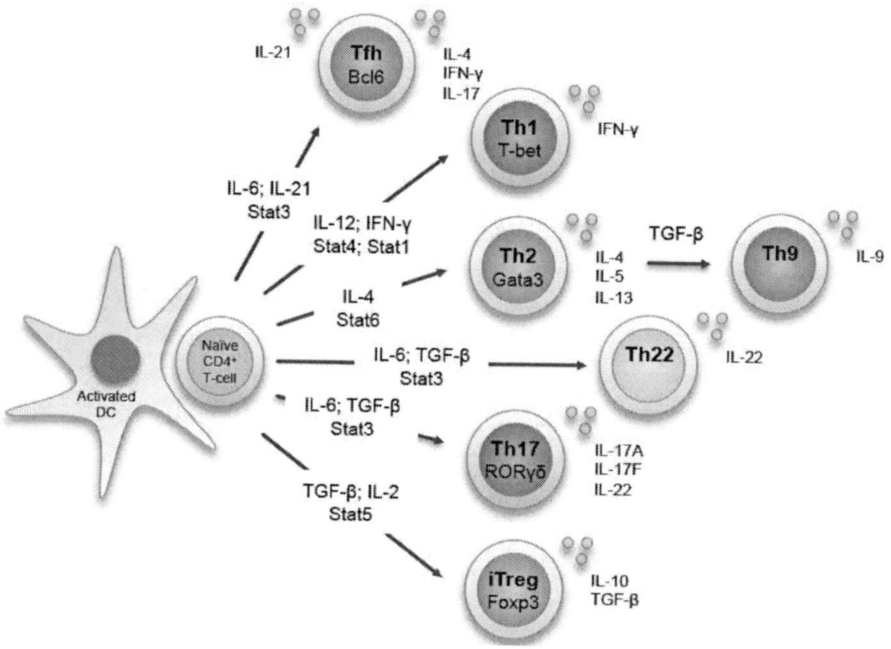


Figure 5. Differentiation of effector $CD4^+$ T-cells subsets in the mouse. During activation in a particular cytokine milieu, naïve $CD4^+$ T-cells may differentiate into one of several lineages of T helper (Th) cells, including Th1, Th2, Th17, and inducible regulatory T-cells (iTregs). Different differentiation programs triggered by TCR and cytokine signaling are mainly based on the activation of Stat proteins and induction of master transcription factors that collaborate in the regulation of unique patterns of cytokine production. However, cytokine production by Th cells might be more flexible than previously though and recently T follicular helper (Tfh), Th9 and Th22 cells have been described. Whether these subsets represent distinct lineages remains to be elucidated.

Association of the IL-2 receptor α chain to the constitutively expressed β and γ chains in resting T cells confers high affinity to IL-2 and allows activated cells to respond to very low concentrations of IL-2. IL-2 signals through activation of Stat5 (signal transducer and activator of transcription 5) proteins to promote T cell proliferation and survival (Moriggl et al., 1999). $CD4^+$ T cell differentiation into different T helper lineages is determined by the cytokine milieu present during the initial proliferative phase of T cell activation (Zhu et al. 2010), whereas $CD8^+$ T cells are already predestined to become cytotoxic T cells (Arens and Schoenberger, 2010; Rutishauser and Kaech 2010). T helper cell differentiation is classically regarded as a dichotomy between two main cell types, termed Th1 and Th2 (O'Shea and Paul, 2010; Zhu et al. 2010) (Figure 5).

Th1 cells produce IFNγ as their signature cytokine and are predominantly involved in cell-mediated immunity against intracellular pathogens, through activation of macrophages and directing the production of opsonising antibodies by B cells (i.e. antibodies that trigger the elimination of pathogens by phagocytosis).

In contrast, Th2 cells fail to produce IFNγ and their signature cytokines are IL-4, IL-5 and IL-13 encoded by genes clustered in a Th2 cytokine locus, also containing the constitutively expressed *Rad50* gene (Figure 6). Th2 cells are the most effective activators of B cell proliferation and antibody production, thereby mediating humoral immunity essential for the eradication of extracellular pathogens.

The differentiation process of naïve $CD4^+$ T cells into distinct T helper effector subsets involves both TCR- and cytokine-mediated signaling (O'Shea and Paul, 2010; Zhu et al. 2010), which are critical for early cytokine production and induction of master transcription factor expression. Cytokine-mediated signaling results in the activation of Stat proteins, which collaborates with master transcription factors in the regulation of distinct patterns of cytokine gene expression, selection of differentiated lineage cells by positive feedback loops and active repression of alternative lineage fates. Two major signaling pathways facilitate Th1 development, one involving IL-12/Stat4 and the other involving IFNγ/Stat1/T-bet. Activated Stat4 can directly induce IFNγ production and expression of IL-12Rβ2 and the T-box transcription factor T-bet during Th1 differentiation (Usui et al., 2003; Usui et al., 2006). Activation of Stat1 by IFNγ is also important for the induction of T-bet during Th1 differentiation (Afkarian et al., 2002; Lighvani et al., 2001).

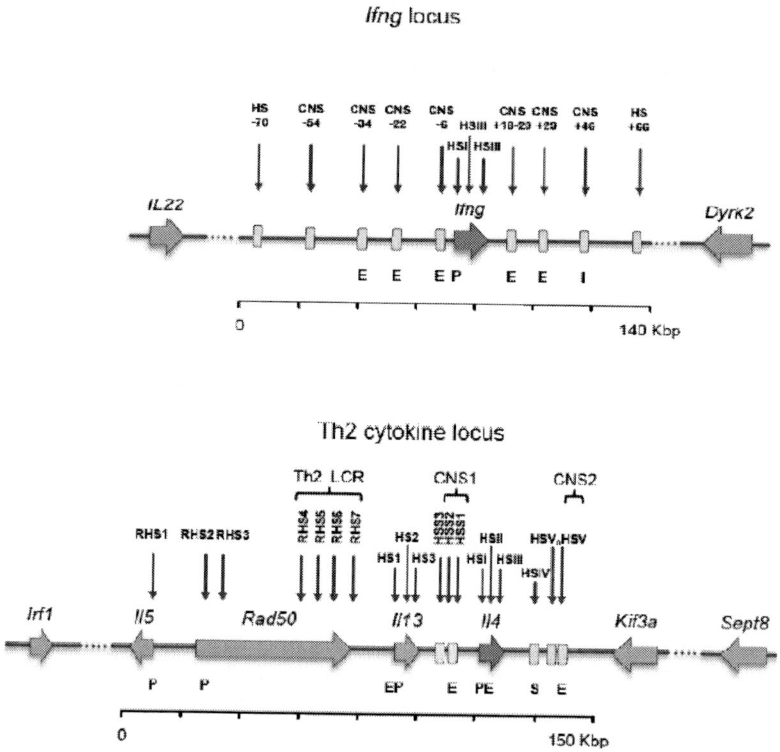


Figure 6. Schematic representation of the mouse *Ifng* and Th2 cytokine loci. The *Ifng* and Th2 cytokine loci span a 140 and 150 Kbp region, respectively. *Cis*-regulatory elements, denoted by the presence of DNaseI hypersensitive sites (HS) and conservation of non-coding sequences (CNS) are indicated by arrows. E, enhancer; P, promoter; I, insulator; LCR, locus control region; S, silencer; RHS, *Rad50* HS.

T-bet is a major transcription factor for inducing IFNγ production and T-bet-deficient $CD4^+$ T cell have severe defects in Th1 cell differentiation (Szabo et al., 2000b; Szabo et al., 2002). T-bet induces IFNγ and IL12Rβ2 expression, thus promoting selective Th1 cell expansion in response to IL-12 (Mullen et al., 2001; Mullen et al., 2002). Th2 cytokine production is dependent on the transcription factor Gata3 (Zhang et al., 1997; Zheng and Flavell, 1997), which is rapidly induced by IL-4 through Stat6 (Kaplan et al., 1996; Kurata et al., 1999; Shimoda et al., 1996; Takeda et al., 1996; Zhu et al., 2001). During the induction phase of Th2 polarization, early TCR-dependent and IL-4/Stat6-independent IL-4 transcription requires induction of Gata3 expression by TCR-signalling and activated Stat5 in response to IL-2 (Yamane et al., 2005). Gata3 expression is further enhanced by IL-4/Stat6 signalling and an auto-regulatory positive feedback loop (Ouyang et al., 2000). In the absence of Gata3 Th2 differentiation is totally abolished (Pai et al., 2004; Zhang et al., 1999; Zhu et al., 2004). In contrast, deletion of *Gata3* from fully differentiated Th2 cells has only a modest effect on IL-4 production but completely blocks the production of IL-5 and IL-13 (Zhu et al., 2004). Importantly, master regulators and Stat proteins collaborate with other transcription factors (for example, Runx1, Runx3, IFN regulatory factor (Irf) 4, Irf1, Gfi-1, Ikaros and c-Maf) and signaling pathways ($Ca^{2+}$/Nfat, NF-κB and Notch signaling) in the fine-tuning of T helper cell differentiation and function (Zhu et al. 2010). Of interest, different Notch ligands on APCs have been proposed to instruct distinct T helper cell fates (Amsen et al., 2004). Particularly, Notch appears to direct Th2 differentiation by directly regulating the expression of *Gata3* and *Il4* (Amsen et al., 2007; Amsen et al., 2004; Tanaka et al., 2006).

More recently, Th17 cells have been described as a distinct T helper subset that does not express T-bet or Gata3 and which differentiation is inhibited by IFNγ or IL-4 (Aggarwal et al., 2003; Cua et al., 2003; Harrington et al., 2005; Murphy et al., 2003; Park et al., 2005) (Figure 5). These cells are characterized by the production of IL-17A, IL-17F and IL-22 as signature cytokines and contribute to host defense against extracellular pathogens particularly at mucosal surfaces. In mice, both TGF-β and IL-6 are required to drive Th17 differentiation through activation Stat3 and induction of the retinoid related orphan receptor-γt (RORγt) (Bettelli et al., 2006; Ivanov et al., 2006; Mangan et al., 2006; Nurieva et al., 2007; Veldhoen et al., 2006; Yang et al., 2007; Zhou et al., 2007). The residual IL-17 production in RORγt-deficient cells appears to be dependent on the activity of a related nuclear receptor RORα, which is also upregulated in Th17 cells (Yang et al., 2008). Th17 cells are also good producers of IL-21 and express IL-23R as part of their differentiation process (Korn et al., 2007; Nurieva et al., 2007; Yang et al., 2007; Zhou et al., 2007). Similarly to IL-6, IL-21 and IL-23 also signal through activation of Stat3 and in Stat3-deficient mice IL-17 production in $CD4^+$ T cells is abolished (Harris et al., 2007; Korn et al., 2007; Laurence and O'Shea, 2007; Mathur et al., 2007; Nurieva et al., 2007; Yang et al., 2007; Zhou et al., 2007). IL-21 and IL-23 possibly function in a positive feedback loop to promote expansion and terminal differentiation of Th17 cells at later stages. Stat4 also mediates IL-23 signals and was also shown to direct the development of Th17 cells (Hildner et al., 2007; Mathur et al., 2007).

Recent studies on T helper cell differentiation have revealed more flexibility in cytokine production than predicted by conventional models of T helper cell lineage commitment (Figure 5). One example of T helper cell phenotype plasticity is that Th2 cells can acquire IL-9-producing capacity in the presence of TGF-β via Irf4 activation, being thereby reprogrammed into a "Th9" pathway (Dardalhon et al., 2008; Staudt et al., 2010; Veldhoen et al., 2008). Similarly, differentiation of T helper cells producing IL-22 ("Th22") and not IL-17 in response to IL-6 and TGF-β has recently been identified and shown to depend on the transcription factors RORγt and aryl hydrocarbon hydrocarbon receptor (AHR) (Duhen et al., 2009; Trifari et al., 2009). T follicular helper (Tfh) are yet another $CD4^+$ T cell population with a particular function, helping B cells make antibody responses to T cell-dependent antigen during a germinal center reaction (see below) (Breitfeld et al., 2000; Schaerli et al., 2000). Tfh cells express Bcl6 as their master transcription factor and may produce Th1, Th2 or Th17 cytokines, depending on the conditions of their initial activation (Johnston et al., 2009; Nurieva et al., 2008; Nurieva et al., 2009; Reinhardt et al., 2009; Vogelzang et al., 2008). Whether Tfh cells are a distinct lineage (essentially parallel to Th1, Th2 or Th17 cells) or a particular phenotypic state of some or all of these effector lineages remains unresolved.

## Regulation of Lineage-Specific Patterns of Cytokine Gene Expression: The Th1/Th2 Paradigm

During T helper cell differentiation, fully polarized T helper cells are generated through lineage commitment and active suppression of alternative lineage fates. The mutual exclusivity among master transcription factors and Stat proteins regulated at the transcriptional levels and through suppressive protein-protein interactions appears to be one of the main mechanisms for cross-regulation during T helper differentiation (O'Shea and Paul, 2010; Zhu et al. 2010). For example, Gata3 suppresses Th1 development by downregulation of Stat4 (Ouyang et al., 1998; Usui et al., 2003) and T-bet restrains Th2 development by inhibiting Gata3 expression (Usui et al., 2006) and physically interacting with Gata3 thereby repressing its function (Hwang et al., 2005).

However, more precise control of gene expression during T helper differentiation is achieved through epigenetic processes, which facilitate heritable and stable programs of gene expression, while preserving the potential for these programs to be modified in response to environmental changes (Ansel et al., 2006; Lee et al., 2006; Wilson et al., 2009). These include chromatin remodelling (induction of DNaseI hypersensitive sites by altering nucleossome positioning) through histone acetylation and/or methylation and DNA demethylation at critical cis-regulatory elements present in cytokine loci (Figure 6).

Changes to high-order chromatin structure at *Ifng* and Th2 cytokine loci, have also been implicated in the coordinate regulation of cytokine gene expression. In naïve CD4$^+$ T cells, low-level expression of Th2 cytokines and IFNγ can be detected upon activation by TCR ligation (Ansel et al., 2004; Grogan et al., 2001), which is supported by both permissive and repressive histone modifications at both loci (Lee et al., 2006; Wilson et al., 2009). This bivalent epigenetic state may help to "poise" the Th2 cytokine and *Ifng* loci for either expression or complete silencing during T helper differentiation.

In the Th2 locus, Th2 cytokine genes promoters and the Th2 locus control region (LCR) come into close spatial proximity to form a higher-order chromatin structure, further suggesting that early expression of the Th2 cytokines in naive T cells is supported by an initial "poised" three-dimensional chromatin configuration (Lee et al., 2005; Spilianakis and Flavell, 2004) (Figure 7).

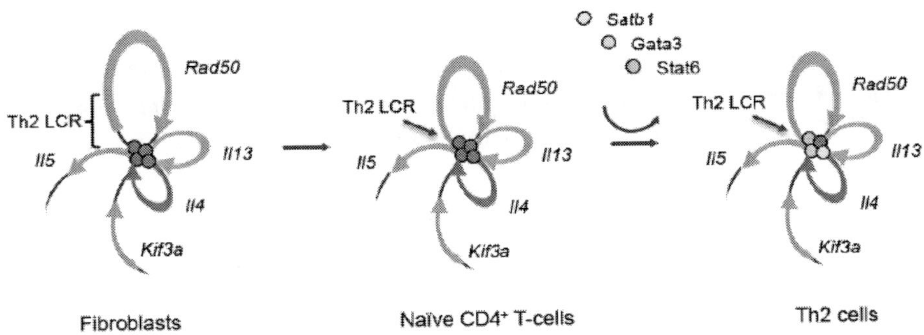


Figure 7. "Poised" chromatin configuration in the Th2 cytokine loci. An initial chromatin configuration is detected in non T-cell lineage cells, such as fibroblasts, characterized by close spatial proximity between Th2 cytokine gene promoters. In naïve CD4$^+$ T-cells, the Th2 locus control region (LCR) further contributes to this higher order chromatin conformation to generate a "poised" state. Upon Th2 differentiation, recruitment of cell-specific factors, such as Gata3, Stat6 or Satb1, to this "poised" chromatin configuration potentiates the coordinate expression of Th2 cytokine genes. Figure adapted from Lee et al., 2005.

Interchromosomal interactions between the *Ifng* and the Th2 cytokine locus have also been described in naïve CD4$^+$ T cells and are apparently lost upon Th1 or Th2 cell differentiation (Spilianakis et al., 2005). However, unequivocal data required to prove that interchromosomal cytokine gene regulation truly represents another level of control within the nucleus is lacking (Williams et al. 2010). Upon Th2 differentiation, a substantial increase in the transcriptional activity of *Il4, Il5 and Il13* and concomitant silencing of *Ifng* are observed.

The converse pattern of gene activation and silencing is present in differentiating Th1 cells. Several studies have demonstrated that cell type-specific transcription factors and Stat proteins binding to cis-regulatory elements of cytokine genes directs subsequent recruitment of chromatin modifying proteins, which initiate substantial and reciprocal alterations in the chromatin structure of the Th2 cytokine and *Ifng* loci (Ansel et al., 2006; Lee et al., 2006; Wilson et al., 2009).

Likewise, both Stat4 and T-bet have been implicated in permissive histone modifications (acetylation and/or methylation) at the *Ifng* locus (Avni et al., 2002; Djuretic et al., 2007; Fields et al., 2002; Mullen et al., 2001; Szabo et al., 2000b; Thieu et al., 2008; Tong et al., 2005; Zhang and Boothby, 2006). Similarly, Stat6 and Gata3 are involved in Th2 locus permissive histone modifications and DNA demethylation (Fields et al., 2002; Hutchins et al., 2002; Lee et al., 2002; Lee et al., 2000; Makar et al., 2003; Ouyang et al., 2000; Takemoto et al., 2000; Wurster and Pazin, 2008; Yamashita et al., 2004). By contrast, Stat6 and Gata3 binding to the *Ifng* locus when naïve $CD4^+$ T cells differentiate into Th2 cells is associated with increased acquisition of repressive histone methylation marks (Chang and Aune, 2007). During Th2 differentiation, increased DNA methylation at the *Ifng* locus is also observed (Chang and Aune, 2007; Jones and Chen, 2006; Schoenborn et al., 2007). In addition, T-bet is required for the silencing of *Il4* in Th1 cells (Ansel et al., 2004; Baguet and Bix, 2004; Djuretic et al., 2007; Naoe et al., 2007). During Th1 and Th2 differentiation, chromatin looping at *Ifng* and Th2 cytokine loci has also proven to be essential for cytokine gene expression. Indeed, both Stat6 and Gata3 are responsible for the establishment and/or maintenance of a “poised” chromatin configuration in the Th2 cytokine locus of developing Th2 cells (Lee et al., 2005; Spilianakis and Flavell, 2004). Further changes to this conformation occur through Th2-specific triggering of Satb1 (special AT-rich binding protein 1, an architectural factor) expression, which promotes the formation of additional loops and more intimate interactions between cis-regulatory elements and the *Il4, Il5 and Il13* promoters (Cai et al., 2006). In the absence of Satb1 these changes are lost and Th2 cytokine expression is compromised (Cai et al., 2006). In Th1 cells, T-bet has also been implicated in the formation of Th1-specific Ifng locus architecture that brings cis-regulatory regions in close proximity to the Ifng promoter (Sekimata et al., 2009). Finally, the expression of T-bet and Gata3 is not essential but does help to maintain the Th1- and Th2-specific gene expression program, respectively (Martins et al., 2005; Mullen et al., 2002; Pai et al., 2004; Yamashita et al., 2004; Zhu et al., 2004). Although cytokine gene expression can be partially sustained through hereditable epigenetic modifications, it benefits from collaboration between lineage-specifying transcription factors and epigenetic processes.

## CTCF Is Required for Proper Gene Regulation at the INFγ and Th2 Cytokine Loci

During T helper cell differentiation, coordinate regulation of cytokine gene expression depends on long-range chromatin loops bringing distal *cis*-regulatory elements into intimate contact with cytokine genes (Cai et al., 2006; Lee et al., 2005; Sekimata et al., 2009; Spilianakis and Flavell, 2004). Both T helper cell lineage-specific transcription factors and Stat proteins have been implicated in the control of these intrachromosomal interactions (Cai et al., 2006; Lee et al., 2005; Sekimata et al., 2009; Spilianakis and Flavell, 2004). However, our current understanding of the mechanisms underlying cell type-specific differences in higher-order chromatin structure at cytokine gene loci and their contribution to gene expression is far from complete.

We have used conditional gene targeting to study the function of CTCF in peripheral T differentiation towards Th1 or Th2 cells (Ribeiro de Almeida et al., 2009). Chromatin immunoprecipitation analysis revealed multiple CTCF binding sites in the Th2 cytokine locus. Conditional deletion of the *Ctcf* gene in double-positive thymocytes allowed development of peripheral T cells, but their activation and proliferation upon anti-CD3/anti-CD28 stimulation in vitro was severely impaired. Nevertheless, when TCR signaling was circumvented with phorbol ester and ionomycin, we observed proliferation of CTCF-deficient T cells, enabling the analysis of Th2 differentiation in vitro. We found that in CTCF-deficient Th2 polarization cultures, transcription of IL-4, IL-5, and IL-13 was strongly reduced (Ribeiro de Almeida et al., 2009). By contrast, CTCF deficiency had a moderate effect on IFNγ production in Th1 cultures and IL-17 production in Th17 cultures was unaffected. Consistent with a Th2 cytokine defect, CTCF-deficient mice had very low levels of IgG1 and IgE in their serum, but IgG2c was close to normal. In CTCF-deficient Th2 cultures, cells were polarized toward the Th2 lineage, as substantiated by induction of the key transcriptional regulators Gata3 and special AT-rich binding protein 1 (Satb1) and down-regulation of T-bet. Also, Stat4 expression was low, indicating that in the absence of CTCF, Gata3 still operated as a negative regulator of Stat4. Taken together, these findings show that CTCF is essential for Gata3- and Satb1-dependent regulation of Th2 cytokine gene expression (Ribeiro de Almeida et al., 2009).

We found ~50% reduction in IFNγ-producing cells when naïve $CD4^+$ T cells from *CD4-Cre Ctcf*$^{f/f}$ mice are cultured in Th1-polarized conditions. The mechanism for reduced IFNγ production in the absence of CTCF has been illuminated by a another recently reported study (Sekimata et al., 2009). This study demonstrates that T-bet promotes *Ifng* expression in part by facilitating CTCF binding and chromatin looping at the *Ifng* locus. Particularly, CTCF acts to establish a Th1-cell specific *Ifng* locus architecture, which helps to drive the juxtaposition of T-bet-binding enhancers and the flanking CTCF-binding elements to *Ifng* and to promote *Ifng* expression (Figure 8).

Another study in human Th1 cells demonstrated that cohesins form the topological and mechanistic basis for CTCF-mediated chromatin looping at the *IFNG* locus (Hadjur et al., 2009). Consistently, we also observed reduced IFNγ production in CTCF-deficient $CD8^+$ T-cell cultures (C.R.A., unpublished). Given that in $CD8^+$ T-cells T-bet acts in concert with other, highly homologous, T-box transcription factor called Eomes to promote IFNγ production (Intlekofer et al., 2005; Pearce et al., 2003; Szabo et al., 2002), it would be interesting to investigate if CTCF cooperates with T-bet and/or Eomes for *Ifng* locus architecture in $CD8^+$ T-cells.

In all vertebrates except rodents, the *IFNG* gene is not clustered with other co-expressed cytokines and the nearest upstream genes are *IL26* and *IL22*, which are mainly expressed by Th17 cells. In rodents, complex structural rearrangements resulted in near complete deletion of *IL26*, except for a distal regulator element located ~70 kb upstream of *Ifng* (Schoenborn et al., 2007).

The mouse -70 kb region exhibits both barrier and enhancer-blocking insulator activity and CTCF occupancy is detected in naïve, Th1 and Th2 $CD4^+$ T-cells (Sekimata et al., 2009). Interestingly, we observed a small population of $IFN\gamma^+$ cells in CTCF-deficient Th17 cultures (Ribeiro de Almeida et al., 2009), suggesting that binding of CTCF to the -70 kb region insulate the *Ifng* locus from the neighboring *Il22* gene. Expression of IFNγ can result in impaired differentiation of Th17 cells (Harrington et al., 2005; Park et al., 2005), although *in vitro* this might be masked by the presence of IFNγ neutralizing antibodies in Th17-polarizing culture conditions.

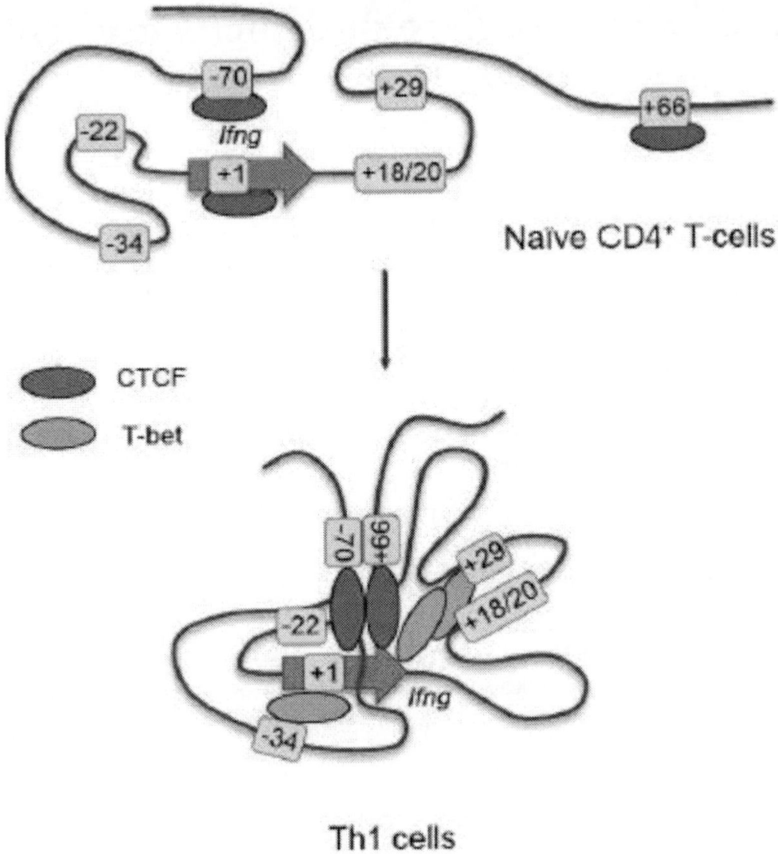


Figure 8. Model for CTCF-mediated higher-order chromatin conformation of the *Ifng* locus. In naive CD4$^+$ T-cells, CTCF is bound primarily at hypersensitive site (HS) -70 and T-bet is not bound. Upon Th1 cell differentiation, T-bet binds to the *Ifng* promoter and to the conserved non-coding sequence (CNS) -34, CNS+18/20, and CNS+29 enhancers, and CTCF binds strongly at HS +1 (in intron 1 of *Ifng*) and at HS +66 in addition to HS -70. This binding contributes to and is required for the juxtaposition of each of these distal regulatory elements to *Ifng* and its promoter. In this active locus conformation, the CNS -34 enhancer is in close proximity to *Ifng* gene. Figure adapted from Sekimata et al., 2009.

It would be interesting to investigate whether differentiation of Th17 cells in *CD4-Cre Ctcf$^{f/f}$* mice is impaired in vivo, for example upon induction of Th17-mediated inflammatory or autoimmune disorders, such as inflammatory bowel disease (IBD), experimental autoimmune encephalomyelitis (EAE) or collagen induced arthritis (CIA). Similarly, whether *Il22* expression is upregulated in CTCF-deficient Th1 cultures and impairs Th1 polarization remains to be tested.

# CTCF, the Th2 Cytokine Locus and Th2 Lineage Commitment

Our findings in conditional CTCF knock-out mice support a critical role for CTCF in the control of Th2 cytokine gene expression (Ribeiro de Almeida et al., 2009). We found that transcription of *Il4*, *Il5* and *Il13* is strongly reduced when naïve $CD4^+$ T cells from *CD4-Cre* $Ctcf^{f/f}$ mice are cultured in Th2-polarizing conditions. Nevertheless, cells were polarized towards the Th2 lineage as substantiated by induction of the key transcriptional regulators Gata3 and Satb1 and low expression of T-bet and Stat4. Based on these findings we propose that CTCF is essential for Gata3- and Satb1-dependent regulation of Th2 cytokine gene expression. However, the mechanisms by which CTCF might facilitate Th2 cytokine production remain largely elusive. As shown in Figure 9, we identified four constitutive CTCF binding sites (CBS) in the mouse Th2 cytokine locus: two sites upstream of *Il5* (CBS-6 and CBS-7) and one downstream of *Il4* within *Kif3a* (CBS-1) flanking the Th2 cytokine locus and one site in the intergenic region between *Il13* and *Il4* (CBS-3) located to the DNaseI hypersensitive site Hss3 (Ribeiro de Almeida et al., 2009), which is of unknown function (Takemoto et al., 2000). Previous studies have shown that both Gata3 and Stat6 are required for the establishment and/or maintenance of a "poised" chromatin configuration in the Th2 cytokine locus, where cytokine gene promoters and the LCR come into close spatial proximity (Lee et al., 2005; Spilianakis and Flavell, 2004).

Further changes to this conformation are induced upon expression of Satb1, which mediates the formation of a more densely looped, transcriptionally active chromatin structure at the Th2 cytokine locus (Cai et al., 2006). One possible explanation for the Th2 cytokine expression defect observed in CTCF-deficient Th2 cells is that in the absence of CTCF the recruitment of key transcription factors for Th2 development, such as Gata3, Stat6, Satb1, c-Maf or Irf4, to the Th2 cytokine locus is impaired (Figure 9). Indeed, one of the CTCF binding sites we identified (CBS-3) is located near to CNS-1, which has been shown to bind the C-terminal Zn-finger of Gata3 and to be critical for Th2 cytokine expression (Mohrs et al., 2001; Takemoto et al., 2000).

Preliminary experiments using ChIP assay suggested that CTCF deficiency in Th2 cells does not result in decreased occupancy of Gata3 at the Th2 cytokine locus in several Gata3 binding sites (C.R.A., unpublished). Another possibility is that CTCF organizes higher-order chromatin conformation at the Th2 cytokine locus required for transcription of *Il4*, *Il5* and *Il13* genes.

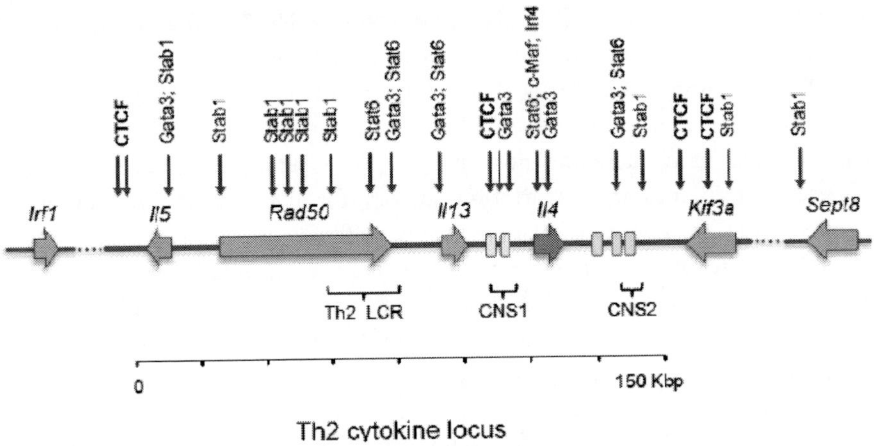


Figure 9. Schematic representation of the binding sites for CTCF and key Th2 transcription factors in the mouse Th2 cytokine locus. CTCF binding sites in the Th2 cytokine locus are interspersed with the binding sites for key Th2 transcription factors such as Gata3, Stat6, Satb1, c-Maf and Irf4. CTCF binding sites (CBS) upstream of *Il5* (CBS-6 and CBS-7) and downstream of *Il4* within *Kif3a* (CBS-1), flank the Th2 cytokine locus. CTCF binding in the intergenic region between *Il13* and *Il4* (CBS-3) locates to the DNaseI hypersensitive site Hss3 near to CNS-1.

This hypothesis can be tested using 3C-type experiments to interrogate Th2 cytokine locus interactions in CTCF-deficient and wild-type naïve, Th1 and Th2 $CD4^+$ T-cells. For instance, upstream and downstream CTCF-binding elements could cooperate with the CTCF-binding element in the *Il13-Il4* intergenic region to help approximate Th2 cytokine gene promoters and *cis*-regulatory elements (such as the LCR or interposed enhancers) during Th2 differentiation. In this context, it would be interesting to investigate if similarly to the *Ifng* locus (Sekimata et al., 2009), Th2 cell-specific CTCF binding and chromatin looping also occurs at the Th2 cytokine locus. It also remains to be tested whether CTCF has a role in the maintenance of a transcriptionally active chromatin structure at Th2 cytokine locus. Further experiments where CTCF is targeted *in vitro* once polarization to the Th2 lineage is complete (for example using RNAi or recombinant soluble Cre approaches) would elucidate this.

During Th2 differentiation Gata3-mediated counter-regulatory mechanisms are required for the suppression of alternative T helper lineage fates. For example, Gata3 suppresses Th1 differentiation by downregulation of *Stat4* (Ouyang et al., 1998; Usui et al., 2003). Gata3 has also been shown to inhibit *Foxp3* expression during iTreg polarization (Mantel et al., 2007). We found that enforced Gata3 expression counteracts Th17 differentiation via (i) induction of IL-4 expression and (ii) downregulation of *Stat3*, *Stat4*, *Nfatc2* and *RORγt* transcription in an IL-4-independent way (van Hamburg et al., 2008). It is therefore conceivable that these transcription factors are direct Gata3 targets. One intriguing possibility to be tested is whether CTCF also cooperates with master transcription factors for the negative regulation of genes essential for alternative T helper lineage fates.

Taken together, CTCF is required for coordinate regulation of *Ifng* and Th2 cytokine gene expression during Th1 and Th2 cell differentiation, respectively. In particular, studies at the *Ifng* locus show that T-bet-dependent CTCF occupancy is on the basis of a Th1-cell specific *Ifng* locus architecture and *Ifng* expression (Sekimata et al., 2009).

Importantly, these findings demonstrate a presumably widespread mechanism of cooperation between a ubiquitous architectural factor (CTCF) and a lineage-specific transcription factor (T-bet) in cell type-specific gene regulation.

## Concluding Remarks

The study of lymphocyte development is important for understanding how lymphocytes acquire their antigen specificity and function, and at the same time constitute an excellent model system to study developmentally regulated gene expression. As for the regulation of T helper cytokine gene expression, the role of CTCF at the *Ifng* locus is parallel to its role in the *MHCII* locus (Figure 10).

In both loci CTCF cooperates with cell type-specific factors (T-bet or CIITA/RFX) for enhancer-promoter long-range interactions and gene expression. It remains to be established if at the Th2 cytokine locus CTCF operates in a similar manner for coordinate regulation of gene expression, involving e.g. Gata3 and Satb1 (Figure 10). An alternative view is that CTCF-based chromatin loops are required for the establishment and/or maintenance of a transcriptionally active chromatin structure at the Th2 cytokine locus.

A similar mechanism has been predicted at the *β-globin* locus, although destabilization of CTCF contacts has no effect on gene expression. Taken together, these studies demonstrate that CTCF-mediated chromatin loops can facilitate communication between *cis*-regulatory elements but can also lead to the exclusion of interactions between them. The spatial positioning of CTCF binding sites with respect to genes and other regulatory elements and the affinity of long-range interactions dictates the types of CTCF-based chromatin loops formed and final outcome in terms of gene regulation.

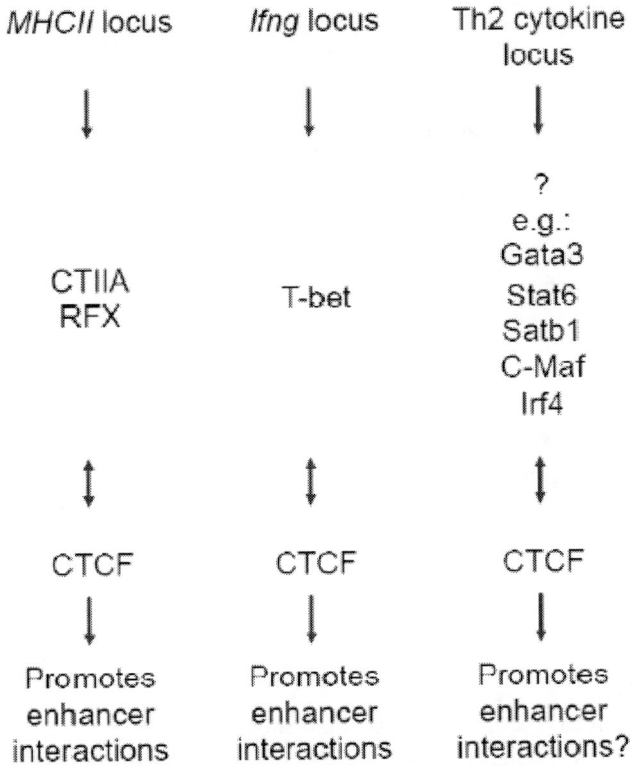


Figure 10. Comparison between CTCF function at several developmentally regulated loci. CTCF-mediated chromatin loops can facilitate the communication between *cis*-regulatory elements such as enhancers and promoters. At the human *MHCII* and mouse *Ifng* loci CTCF has been shown to cooperate with cell-type specific transcription factors (CIITA/RFX and T-bet) for long-range interactions between distal cis-regulatory elements. We hypothesize that a similar mechanism accounts for CTCF function – to promote enhancer interactions - at the Th2 locus.

Dissecting the molecular mechanisms underlying the generation of a wide repertoire of antigen receptors and the expression of effector cytokines may (i) facilitate the development of clinical strategies for the generation of a functional lymphoid compartment, able to fight a vast array of potential pathogens while maintaining tolerance to self (for example, in the context of HSC transplantation) and, (ii) contribute to understand the etiology of Th2-mediated diseases (such as, asthma and allergy) and improve current therapies. Despite CTCF being a ubiquitously expressed transcription factor, the possibility of topical treatment with CTCF-inhibitors for pulmonary Th2 diseases such as asthma is a realistic one, paralleling e.g. inhaled corticosteroids.

## Acknowledgments

This work was supported by Fundação para a Ciência e a Tecnologia (C. R. A.) and the Dutch Cancer Foundation (KWF/NKB, R.W.H.).

## References

Afkarian, M., Sedy, J.R., Yang, J., Jacobson, N.G., Cereb, N., Yang, S.Y., Murphy, T.L., and Murphy, K.M. (2002). T-bet is a STAT1-induced regulator of IL-12R expression in naive CD4+ T cells. *Nat. Immunol. 3*, 549-557.

Aggarwal, S., Ghilardi, N., Xie, M.H., de Sauvage, F.J., and Gurney, A.L. (2003). Interleukin-23 promotes a distinct CD4 T cell activation state characterized by the production of interleukin-17. *J. Biol. Chem. 278*, 1910-1914.

Amsen, D., Antov, A., Jankovic, D., Sher, A., Radtke, F., Souabni, A., Busslinger, M., McCright, B., Gridley, T., and Flavell, R.A. (2007). Direct regulation of Gata3 expression determines the T helper differentiation potential of Notch. *Immunity, 27*, 89-99.

Amsen, D., Blander, J.M., Lee, G.R., Tanigaki, K., Honjo, T., and Flavell, R.A. (2004). Instruction of distinct CD4 T helper cell fates by different notch ligands on antigen-presenting cells. *Cell 117*, 515-526.

Ansel, K.M., Djuretic, I., Tanasa, B., and Rao, A. (2006). Regulation of Th2 differentiation and Il4 locus accessibility. *Annu. Rev. Immunol. 24*, 607-656.

Ansel, K.M., Greenwald, R.J., Agarwal, S., Bassing, C.H., Monticelli, S., Interlandi, J., Djuretic, I.M., Lee, D.U., Sharpe, A.H., Alt, F.W., *et al.*

(2004). Deletion of a conserved Il4 silencer impairs T helper type 1-mediated immunity. *Nat. Immunol. 5*, 1251-1259.

Arens, R., and Schoenberger, S.P. (2010) Plasticity in programming of effector and memory CD8 T-cell formation. *Immunol. Rev. 235*, 190-205.

Avni, O., Lee, D., Macian, F., Szabo, S.J., Glimcher, L.H., and Rao, A. (2002). T(H) cell differentiation is accompanied by dynamic changes in histone acetylation of cytokine genes. *Nat. Immunol. 3*, 643-651.

Baguet, A., and Bix, M. (2004). Chromatin landscape dynamics of the Il4-Il13 locus during T helper 1 and 2 development. *Proc. Natl. Acad. Sci. U S A, 101*, 11410-11415.

Balciunaite, G., Ceredig, R., Fehling, H.J., Zuniga-Pflucker, J.C., and Rolink, A.G. (2005). The role of Notch and IL-7 signaling in early thymocyte proliferation and differentiation. *European journal of immunology 35*, 1292-1300.

Barski, A., Cuddapah, S., Cui, K., Roh, T.Y., Schones, D.E., Wang, Z., Wei, G., Chepelev, I., and Zhao, K. (2007). High-resolution profiling of histone methylations in the human genome. *Cell, 129*, 823-837.

Bell, A.C., and Felsenfeld, G. (2000). Methylation of a CTCF-dependent boundary controls imprinted expression of the Igf2 gene. *Nature, 405*, 482-485.

Bell, A.C., West, A.G., and Felsenfeld, G. (1999). The protein CTCF is required for the enhancer blocking activity of vertebrate insulators. *Cell, 98*, 387-396.

Bell, J.J., and Bhandoola, A. (2008). The earliest thymic progenitors for T cells possess myeloid lineage potential. *Nature, 452*, 764-767.

Bender, M.A., Byron, R., Ragoczy, T., Telling, A., Bulger, M., and Groudine, M. (2006). Flanking HS-62.5 and 3'HS1, and regions upstream of the LCR are not required for {beta}-globin transcription. *Blood.*

Bender, M.A., Reik, A., Close, J., Telling, A., Epner, E., Fiering, S., Hardison, R., and Groudine, M. (1998). Description and targeted deletion of 5' hypersensitive site 5 and 6 of the mouse beta-globin locus control region. *Blood, 92*, 4394-4403.

Benz, C., Martins, V.C., Radtke, F., and Bleul, C.C. (2008). The stream of precursors that colonizes the thymus proceeds selectively through the early T lineage precursor stage of T cell development. *J. Exp. Med. 205*, 1187-1199.

Bettelli, E., Carrier, Y., Gao, W., Korn, T., Strom, T.B., Oukka, M., Weiner, H.L., and Kuchroo, V.K. (2006). Reciprocal developmental pathways for the generation of pathogenic effector TH17 and regulatory T cells. *Nature, 441*, 235-238.

Breitfeld, D., Ohl, L., Kremmer, E., Ellwart, J., Sallusto, F., Lipp, M., and Forster, R. (2000). Follicular B helper T cells express CXC chemokine receptor 5, localize to B cell follicles, and support immunoglobulin production. *J. Exp. Med. 192*, 1545-1552.

Bulger, M., Schubeler, D., Bender, M.A., Hamilton, J., Farrell, C.M., Hardison, R.C., and Groudine, M. (2003). A complex chromatin landscape revealed by patterns of nuclease sensitivity and histone modification within the mouse beta-globin locus. *Mol. Cell Biol. 23*, 5234-5244.

Cai, S., Lee, C.C., and Kohwi-Shigematsu, T. (2006). SATB1 packages densely looped, transcriptionally active chromatin for coordinated expression of cytokine genes. *Nat. Genet. 38*, 1278-1288.

Chang, S., and Aune, T.M. (2007). Dynamic changes in histone-methylation 'marks' across the locus encoding interferon-gamma during the differentiation of T helper type 2 cells. *Nat. Immunol. 8*, 723-731.

Chen, X., Xu, H., Yuan, P., Fang, F., Huss, M., Vega, V.B., Wong, E., Orlov, Y.L., Zhang, W., Jiang, J., *et al.* (2008). Integration of external signaling pathways with the core transcriptional network in embryonic stem cells. *Cell, 133*, 1106-1117.

Chernukhin, I., Shamsuddin, S., Kang, S.Y., Bergstrom, R., Kwon, Y.W., Yu, W., Whitehead, J., Mukhopadhyay, R., Docquier, F., Farrar, D., *et al.* (2007). CTCF interacts with and recruits the largest subunit of RNA polymerase II to CTCF target sites genome-wide. *Mol. Cell Biol. 27*, 1631-1648.

Chernukhin, I.V., Shamsuddin, S., Robinson, A.F., Carne, A.F., Paul, A., El-Kady, A.I., Lobanenkov, V.V., and Klenova, E.M. (2000). Physical and functional interaction between two pluripotent proteins, the Y-box DNA/RNA-binding factor, YB-1, and the multivalent zinc finger factor, CTCF. *J. Biol. Chem. 275*, 29915-29921.

Chi, A.W., Bell, J.J., Zlotoff, D.A., and Bhandoola, A. (2009). Untangling the T branch of the hematopoiesis tree. *Curr. Opin. Immunol. 21*, 121-126.

Ciofani, M., and Zuniga-Pflucker, J.C. (2005). Notch promotes survival of pre-T cells at the beta-selection checkpoint by regulating cellular metabolism. *Nat. Immunol. 6*, 881-888.

Cua, D.J., Sherlock, J., Chen, Y., Murphy, C.A., Joyce, B., Seymour, B., Lucian, L., To, W., Kwan, S., Churakova, T., *et al.* (2003). Interleukin-23 rather than interleukin-12 is the critical cytokine for autoimmune inflammation of the brain. *Nature, 421*, 744-748.

Cuddapah, S., Jothi, R., Schones, D.E., Roh, T.Y., Cui, K., and Zhao, K. (2009). Global analysis of the insulator binding protein CTCF in chromatin barrier regions reveals demarcation of active and repressive domains. *Genome Res. 19*, 24-32.

Dardalhon, V., Awasthi, A., Kwon, H., Galileos, G., Gao, W., Sobel, R.A., Mitsdoerffer, M., Strom, T.B., Elyaman, W., Ho, I.C., *et al.* (2008). IL-4 inhibits TGF-beta-induced Foxp3+ T cells and, together with TGF-beta, generates IL-9+ IL-10+ Foxp3(-) effector T cells. *Nat. Immunol. 9*, 1347-1355.

Defossez, P.A., Kelly, K.F., Filion, G.J., Perez-Torrado, R., Magdinier, F., Menoni, H., Nordgaard, C.L., Daniel, J.M., and Gilson, E. (2005). The human enhancer blocker CTC-binding factor interacts with the transcription factor Kaiso. *J. Biol. Chem. 280*, 43017-43023.

Djuretic, I.M., Levanon, D., Negreanu, V., Groner, Y., Rao, A., and Ansel, K.M. (2007). Transcription factors T-bet and Runx3 cooperate to activate Ifng and silence Il4 in T helper type 1 cells. *Nat. Immunol. 8*, 145-153.

Donohoe, M.E., Zhang, L.F., Xu, N., Shi, Y., and Lee, J.T. (2007). Identification of a Ctcf cofactor, Yy1, for the X chromosome binary switch. *Mol. Cell, 25*, 43-56.

Duhen, T., Geiger, R., Jarrossay, D., Lanzavecchia, A., and Sallusto, F. (2009). Production of interleukin 22 but not interleukin 17 by a subset of human skin-homing memory T cells. *Nat. Immunol. 10*, 857-863.

Farrell, C.M., West, A.G., and Felsenfeld, G. (2002). Conserved CTCF insulator elements flank the mouse and human beta-globin loci. *Mol. Cell Biol. 22*, 3820-3831.

Fields, P.E., Kim, S.T., and Flavell, R.A. (2002). Cutting edge: changes in histone acetylation at the IL-4 and IFN-gamma loci accompany Th1/Th2 differentiation. *J. Immunol. 169*, 647-650.

Filippova, G.N., Cheng, M.K., Moore, J.M., Truong, J.P., Hu, Y.J., Nguyen, D.K., Tsuchiya, K.D., and Disteche, C.M. (2005). Boundaries between chromosomal domains of X inactivation and escape bind CTCF and lack CpG methylation during early development. *Dev. Cell, 8*, 31-42.

Filippova, G.N., Fagerlie, S., Klenova, E.M., Myers, C., Dehner, Y., Goodwin, G., Neiman, P.E., Collins, S.J., and Lobanenkov, V.V. (1996). An exceptionally conserved transcriptional repressor, CTCF, employs different combinations of zinc fingers to bind diverged promoter sequences of avian and mammalian c-myc oncogenes. *Mol. Cell Biol. 16*, 2802-2813.

Filippova, G.N., Thienes, C.P., Penn, B.H., Cho, D.H., Hu, Y.J., Moore, J.M., Klesert, T.R., Lobanenkov, V.V., and Tapscott, S.J. (2001). CTCF-binding sites flank CTG/CAG repeats and form a methylation-sensitive insulator at the DM1 locus. *Nat. Genet. 28*, 335-343.

Garrett, F.E., Emelyanov, A.V., Sepulveda, M.A., Flanagan, P., Volpi, S., Li, F., Loukinov, D., Eckhardt, L.A., Lobanenkov, V.V., and Birshtein, B.K. (2005). Chromatin architecture near a potential 3' end of the igh locus involves modular regulation of histone modifications during B-Cell development and in vivo occupancy at CTCF sites. *Mol. Cell Biol. 25*, 1511-1525.

Godfrey, D.I., Kennedy, J., Suda, T., and Zlotnik, A. (1993). A developmental pathway involving four phenotypically and functionally distinct subsets of CD3-CD4-CD8- triple-negative adult mouse thymocytes defined by CD44 and CD25 expression. *J. Immunol. 150*, 4244-4252.

Gombert, W.M., Farris, S.D., Rubio, E.D., Morey-Rosler, K.M., Schubach, W.H., and Krumm, A. (2003). The c-myc insulator element and matrix attachment regions define the c-myc chromosomal domain. *Mol. Cell Biol. 23*, 9338-9348.

Gombert, W.M., and Krumm, A. (2009). Targeted deletion of multiple CTCF-binding elements in the human C-MYC gene reveals a requirement for CTCF in C-MYC expression. *PLoS One, 4*, e6109.

Gomos-Klein, J., Harrow, F., Alarcon, J., and Ortiz, B.D. (2007). CTCF-independent, but not CTCF-dependent, elements significantly contribute to TCR-alpha locus control region activity. *J. Immunol. 179*, 1088-1095.

Grogan, J.L., Mohrs, M., Harmon, B., Lacy, D.A., Sedat, J.W., and Locksley, R.M. (2001). Early transcription and silencing of cytokine genes underlie polarization of T helper cell subsets. *Immunity, 14*, 205-215.

Guelen, L., Pagie, L., Brasset, E., Meuleman, W., Faza, M.B., Talhout, W., Eussen, B.H., de Klein, A., Wessels, L., de Laat, W., *et al.* (2008). Domain organization of human chromosomes revealed by mapping of nuclear lamina interactions. *Nature, 453*, 948-951.

Hadjur, S., Williams, L.M., Ryan, N.K., Cobb, B.S., Sexton, T., Fraser, P., Fisher, A.G., and Merkenschlager, M. (2009). Cohesins form chromosomal cis-interactions at the developmentally regulated IFNG locus. *Nature, 460*, 410-413.

Haks, M.C., Lefebvre, J.M., Lauritsen, J.P., Carleton, M., Rhodes, M., Miyazaki, T., Kappes, D.J., and Wiest, D.L. (2005). Attenuation of gammadeltaTCR signaling efficiently diverts thymocytes to the alphabeta lineage. *Immunity, 22*, 595-606.

Hark, A.T., Schoenherr, C.J., Katz, D.J., Ingram, R.S., Levorse, J.M., and Tilghman, S.M. (2000). CTCF mediates methylation-sensitive enhancer-blocking activity at the H19/Igf2 locus. *Nature, 405*, 486-489.

Harrington, L.E., Hatton, R.D., Mangan, P.R., Turner, H., Murphy, T.L., Murphy, K.M., and Weaver, C.T. (2005). Interleukin 17-producing CD4+ effector T cells develop via a lineage distinct from the T helper type 1 and 2 lineages. *Nat. Immunol. 6*, 1123-1132.

Harris, T.J., Grosso, J.F., Yen, H.R., Xin, H., Kortylewski, M., Albesiano, E., Hipkiss, E.L., Getnet, D., Goldberg, M.V., Maris, C.H., *et al.* (2007). Cutting edge: An in vivo requirement for STAT3 signaling in TH17 development and TH17-dependent autoimmunity. *J. Immunol. 179*, 4313-4317.

Hayes, S.M., Li, L., and Love, P.E. (2005). TCR signal strength influences alphabeta/gammadelta lineage fate. *Immunity, 22*, 583-593.

He, X., and Kappes, D.J. (2006). CD4/CD8 lineage commitment: light at the end of the tunnel? *Curr. Opin. Immunol. 18*, 135-142.

Heath, H., Ribeiro de Almeida, C., Sleutels, F., Dingjan, G., van de Nobelen, S., Jonkers, I., Ling, K.W., Gribnau, J., Renkawitz, R., Grosveld, F., *et al.* (2008). CTCF regulates cell cycle progression of alphabeta T cells in the thymus. *Embo J. 27*, 2839-2850.

Hildner, K.M., Schirmacher, P., Atreya, I., Dittmayer, M., Bartsch, B., Galle, P.R., Wirtz, S., and Neurath, M.F. (2007). Targeting of the transcription factor STAT4 by antisense phosphorothioate oligonucleotides suppresses collagen-induced arthritis. *J. Immunol. 178*, 3427-3436.

Ho, I.C., Vorhees, P., Marin, N., Oakley, B.K., Tsai, S.F., Orkin, S.H., and Leiden, J.M. (1991). Human GATA-3: a lineage-restricted transcription factor that regulates the expression of the T cell receptor alpha gene. *Embo J. 10*, 1187-1192.

Hoffman, E.S., Passoni, L., Crompton, T., Leu, T.M., Schatz, D.G., Koff, A., Owen, M.J., and Hayday, A.C. (1996). Productive T-cell receptor beta-chain gene rearrangement: coincident regulation of cell cycle and clonality during development in vivo. *Genes Dev. 10*, 948-962.

Hou, C., Zhao, H., Tanimoto, K., and Dean, A. (2008). CTCF-dependent enhancer-blocking by alternative chromatin loop formation. *Proc. Natl. Acad. Sci. U S A, 105*, 20398-20403.

Hutchins, A.S., Mullen, A.C., Lee, H.W., Sykes, K.J., High, F.A., Hendrich, B.D., Bird, A.P., and Reiner, S.L. (2002). Gene silencing quantitatively controls the function of a developmental trans-activator. *Mol. Cell, 10*, 81-91.

Hwang, E.S., Szabo, S.J., Schwartzberg, P.L., and Glimcher, L.H. (2005). T helper cell fate specified by kinase-mediated interaction of T-bet with GATA-3. *Science, 307*, 430-433.

Intlekofer, A.M., Takemoto, N., Wherry, E.J., Longworth, S.A., Northrup, J.T., Palanivel, V.R., Mullen, A.C., Gasink, C.R., Kaech, S.M., Miller, J.D., *et al.* (2005). Effector and memory CD8+ T cell fate coupled by T-bet and eomesodermin. *Nat. Immunol. 6*, 1236-1244.

Ishihara, K., Oshimura, M., and Nakao, M. (2006). CTCF-dependent chromatin insulator is linked to epigenetic remodeling. *Mol. Cell, 23*, 733-742.

Ivanov, II, McKenzie, B.S., Zhou, L., Tadokoro, C.E., Lepelley, A., Lafaille, J.J., Cua, D.J., and Littman, D.R. (2006). The orphan nuclear receptor RORgammat directs the differentiation program of proinflammatory IL-17+ T helper cells. *Cell, 126*, 1121-1133.

Jhunjhunwala, S., van Zelm, M.C., Peak, M.M., and Murre, C. (2009). Chromatin architecture and the generation of antigen receptor diversity. *Cell, 138*, 435-448.

Johnston, R.J., Poholek, A.C., DiToro, D., Yusuf, I., Eto, D., Barnett, B., Dent, A.L., Craft, J., and Crotty, S. (2009). Bcl6 and Blimp-1 are reciprocal and antagonistic regulators of T follicular helper cell differentiation. *Science, 325*, 1006-1010.

Jones, B., and Chen, J. (2006). Inhibition of IFN-gamma transcription by site-specific methylation during T helper cell development. *Embo J. 25*, 2443-2452.

Jung, D., Giallourakis, C., Mostoslavsky, R., and Alt, F.W. (2006). Mechanism and control of V(D)J recombination at the immunoglobulin heavy chain locus. *Annu. Rev. Immunol. 24*, 541-570.

Kanduri, C., Pant, V., Loukinov, D., Pugacheva, E., Qi, C.F., Wolffe, A., Ohlsson, R., and Lobanenkov, V.V. (2000). Functional association of CTCF with the insulator upstream of the H19 gene is parent of origin-specific and methylation-sensitive. *Curr. Biol. 10*, 853-856.

Kaplan, M.H., Schindler, U., Smiley, S.T., and Grusby, M.J. (1996). Stat6 is required for mediating responses to IL-4 and for development of Th2 cells. *Immunity, 4*, 313-319.

Kim, T.H., Abdullaev, Z.K., Smith, A.D., Ching, K.A., Loukinov, D.I., Green, R.D., Zhang, M.Q., Lobanenkov, V.V., and Ren, B. (2007). Analysis of the vertebrate insulator protein CTCF-binding sites in the human genome. *Cell, 128*, 1231-1245.

Korn, T., Bettelli, E., Gao, W., Awasthi, A., Jager, A., Strom, T.B., Oukka, M., and Kuchroo, V.K. (2007). IL-21 initiates an alternative pathway to induce proinflammatory T(H)17 cells. *Nature, 448*, 484-487.

Krangel, M.S. (2009). Mechanics of T cell receptor gene rearrangement. *Curr. Opin. Immunol. 21*, 133-139.

Kreslavsky, T., Garbe, A.I., Krueger, A., and von Boehmer, H. (2008). T cell receptor-instructed alphabeta versus gammadelta lineage commitment revealed by single-cell analysis. *J. Exp. Med. 205*, 1173-1186.

Kurata, H., Lee, H.J., O'Garra, A., and Arai, N. (1999). Ectopic expression of activated Stat6 induces the expression of Th2-specific cytokines and transcription factors in developing Th1 cells. *Immunity, 11*, 677-688.

Kurukuti, S., Tiwari, V.K., Tavoosidana, G., Pugacheva, E., Murrell, A., Zhao, Z., Lobanenkov, V., Reik, W., and Ohlsson, R. (2006). CTCF binding at the H19 imprinting control region mediates maternally inherited higher-order chromatin conformation to restrict enhancer access to Igf2. *Proc. Natl. Acad. Sci. U S A, 103*, 10684-10689.

Lambrecht, B.N., and Hammad, H. (2010).The role of dendritic and epithelial cells as master regulators of allergic airway inflammation. *Lancet, 376*, 835-843.

Laurence, A., and O'Shea, J.J. (2007). T(H)-17 differentiation: of mice and men. *Nat. Immunol. 8*, 903-905.

Lee, D.U., Agarwal, S., and Rao, A. (2002). Th2 lineage commitment and efficient IL-4 production involves extended demethylation of the IL-4 gene. *Immunity, 16*, 649-660.

Lee, G.R., Kim, S.T., Spilianakis, C.G., Fields, P.E., and Flavell, R.A. (2006). T helper cell differentiation: regulation by cis elements and epigenetics. *Immunity, 24*, 369-379.

Lee, G.R., Spilianakis, C.G., and Flavell, R.A. (2005). Hypersensitive site 7 of the TH2 locus control region is essential for expressing TH2 cytokine genes and for long-range intrachromosomal interactions. *Nat. Immunol. 6*, 42-48.

Lee, H.J., Takemoto, N., Kurata, H., Kamogawa, Y., Miyatake, S., O'Garra, A., and Arai, N. (2000). GATA-3 induces T helper cell type 2 (Th2) cytokine expression and chromatin remodeling in committed Th1 cells. *J. Exp. Med. 192*, 105-115.

Li, T., Hu, J.F., Qiu, X., Ling, J., Chen, H., Wang, S., Hou, A., Vu, T.H., and Hoffman, A.R. (2008). CTCF regulates allelic expression of Igf2 by orchestrating a promoter-polycomb repressive complex 2 intrachromosomal loop. *Mol. Cell Biol. 28*, 6473-6482.

Lighvani, A.A., Frucht, D.M., Jankovic, D., Yamane, H., Aliberti, J., Hissong, B.D., Nguyen, B.V., Gadina, M., Sher, A., Paul, W.E., *et al.* (2001). T-bet is rapidly induced by interferon-gamma in lymphoid and myeloid cells. *Proc. Natl. Acad. Sci. U S A, 98*, 15137-15142.

Ling, J.Q., Li, T., Hu, J.F., Vu, T.H., Chen, H.L., Qiu, X.W., Cherry, A.M., and Hoffman, A.R. (2006). CTCF mediates interchromosomal colocalization between Igf2/H19 and Wsb1/Nf1. *Science, 312*, 269-272.

Lutz, M., Burke, L.J., Barreto, G., Goeman, F., Greb, H., Arnold, R., Schultheiss, H., Brehm, A., Kouzarides, T., Lobanenkov, V., *et al.* (2000). Transcriptional repression by the insulator protein CTCF involves histone deacetylases. *Nucleic Acids Res. 28*, 1707-1713.

Magdinier, F., Yusufzai, T.M., and Felsenfeld, G. (2004). Both CTCF-dependent and -independent insulators are found between the mouse T cell receptor alpha and Dad1 genes. *J. Biol. Chem. 279*, 25381-25389.

Majumder, P., Gomez, J.A., Chadwick, B.P., and Boss, J.M. (2008). The insulator factor CTCF controls MHC class II gene expression and is required for the formation of long-distance chromatin interactions. *J. Exp. Med. 205*, 785-798.

Makar, K.W., Perez-Melgosa, M., Shnyreva, M., Weaver, W.M., Fitzpatrick, D.R., and Wilson, C.B. (2003). Active recruitment of DNA methyltransferases regulates interleukin 4 in thymocytes and T cells. *Nat. Immunol. 4*, 1183-1190.

Mangan, P.R., Harrington, L.E., O'Quinn, D.B., Helms, W.S., Bullard, D.C., Elson, C.O., Hatton, R.D., Wahl, S.M., Schoeb, T.R., and Weaver, C.T. (2006). Transforming growth factor-beta induces development of the T(H)17 lineage. *Nature, 441*, 231-234.

Mantel, P.Y., Kuipers, H., Boyman, O., Rhyner, C., Ouaked, N., Ruckert, B., Karagiannidis, C., Lambrecht, B.N., Hendriks, R.W., Crameri, R., *et al.* (2007). GATA3-driven Th2 responses inhibit TGF-beta1-induced FOXP3 expression and the formation of regulatory T cells. *PLoS Biol. 5*, e329.

Marine, J., and Winoto, A. (1991). The human enhancer-binding protein Gata3 binds to several T-cell receptor regulatory elements. *Proc. Natl. Acad. Sci. U S A, 88*, 7284-7288.

Martins, G.A., Hutchins, A.S., and Reiner, S.L. (2005). Transcriptional activators of helper T cell fate are required for establishment but not maintenance of signature cytokine expression. *J. Immunol. 175*, 5981-5985.

Massa, S., Balciunaite, G., Ceredig, R., and Rolink, A.G. (2006). Critical role for c-kit (CD117) in T cell lineage commitment and early thymocyte development in vitro. *European journal of immunology, 36*, 526-532.

Mathur, A.N., Chang, H.C., Zisoulis, D.G., Stritesky, G.L., Yu, Q., O'Malley, J.T., Kapur, R., Levy, D.E., Kansas, G.S., and Kaplan, M.H. (2007). Stat3 and Stat4 direct development of IL-17-secreting Th cells. *J. Immunol. 178*, 4901-4907.

Meier, N., Krpic, S., Rodriguez, P., Strouboulis, J., Monti, M., Krijgsveld, J., Gering, M., Patient, R., Hostert, A., and Grosveld, F. (2006). Novel binding partners of Ldb1 are required for haematopoietic development. *Development, 133*, 4913-4923.

Miosge, L., and Zamoyska, R. (2007). Signalling in T-cell development: is it all location, location, location? *Curr. Opin. Immunol. 19*, 194-199.

Mohrs, M., Blankespoor, C.M., Wang, Z.E., Loots, G.G., Afzal, V., Hadeiba, H., Shinkai, K., Rubin, E.M., and Locksley, R.M. (2001). Deletion of a coordinate regulator of type 2 cytokine expression in mice. *Nat. Immunol. 2*, 842-847.

Moriggl, R., Sexl, V., Piekorz, R., Topham, D., and Ihle, J.N. (1999). Stat5 activation is uniquely associated with cytokine signaling in peripheral T cells. *Immunity, 11*, 225-230.

Mullen, A.C., High, F.A., Hutchins, A.S., Lee, H.W., Villarino, A.V., Livingston, D.M., Kung, A.L., Cereb, N., Yao, T.P., Yang, S.Y., *et al.* (2001). Role of T-bet in commitment of TH1 cells before IL-12-dependent selection. *Science, 292*, 1907-1910.

Mullen, A.C., Hutchins, A.S., High, F.A., Lee, H.W., Sykes, K.J., Chodosh, L.A., and Reiner, S.L. (2002). Hlx is induced by and genetically interacts with T-bet to promote heritable T(H)1 gene induction. *Nat. Immunol. 3*, 652-658.

Murphy, C.A., Langrish, C.L., Chen, Y., Blumenschein, W., McClanahan, T., Kastelein, R.A., Sedgwick, J.D., and Cua, D.J. (2003). Divergent pro- and antiinflammatory roles for IL-23 and IL-12 in joint autoimmune inflammation. *J. Exp. Med. 198*, 1951-1957.

Murrell, A., Heeson, S., and Reik, W. (2004). Interaction between differentially methylated regions partitions the imprinted genes Igf2 and H19 into parent-specific chromatin loops. *Nat. Genet. 36*, 889-893.

Naoe, Y., Setoguchi, R., Akiyama, K., Muroi, S., Kuroda, M., Hatam, F., Littman, D.R., and Taniuchi, I. (2007). Repression of interleukin-4 in T helper type 1 cells by Runx/Cbf beta binding to the Il4 silencer. *J. Exp. Med. 204*, 1749-1755.

Nurieva, R., Yang, X.O., Martinez, G., Zhang, Y., Panopoulos, A.D., Ma, L., Schluns, K., Tian, Q., Watowich, S.S., Jetten, A.M., *et al.* (2007). Essential autocrine regulation by IL-21 in the generation of inflammatory T cells. *Nature, 448*, 480-483.

Nurieva, R.I., Chung, Y., Hwang, D., Yang, X.O., Kang, H.S., Ma, L., Wang, Y.H., Watowich, S.S., Jetten, A.M., Tian, Q., *et al.* (2008). Generation of T follicular helper cells is mediated by interleukin-21 but independent of T helper 1, 2, or 17 cell lineages. *Immunity, 29*, 138-149.

Nurieva, R.I., Chung, Y., Martinez, G.J., Yang, X.O., Tanaka, S., Matskevitch, T.D., Wang, Y.H., and Dong, C. (2009). Bcl6 mediates the development of T follicular helper cells. *Science, 325*, 1001-1005.

O'Shea, J.J., and Paul, W.E (2010). Mechanisms underlying lineage commitment and plasticity of helper CD4+ T cells. *Science, 327*, 1098-1102.

Ohlsson, R., Renkawitz, R., and Lobanenkov, V. (2001). CTCF is a uniquely versatile transcription regulator linked to epigenetics and disease. *Trends Genet. 17*, 520-527.

Ouyang, W., Lohning, M., Gao, Z., Assenmacher, M., Ranganath, S., Radbruch, A., and Murphy, K.M. (2000). Stat6-independent GATA-3 autoactivation directs IL-4-independent Th2 development and commitment. *Immunity, 12*, 27-37.

Ouyang, W., Ranganath, S.H., Weindel, K., Bhattacharya, D., Murphy, T.L., Sha, W.C., and Murphy, K.M. (1998). Inhibition of Th1 development mediated by GATA-3 through an IL-4-independent mechanism. *Immunity, 9*, 745-755.

Pai, S.Y., Truitt, M.L., and Ho, I.C. (2004). GATA-3 deficiency abrogates the development and maintenance of T helper type 2 cells. *Proc. Natl. Acad. Sci. U S A, 101*, 1993-1998.

Palstra, R.J., Tolhuis, B., Splinter, E., Nijmeijer, R., Grosveld, F., and de Laat, W. (2003). The beta-globin nuclear compartment in development and erythroid differentiation. *Nat. Genet. 35*, 190-194.

Parelho, V., Hadjur, S., Spivakov, M., Leleu, M., Sauer, S., Gregson, H.C., Jarmuz, A., Canzonetta, C., Webster, Z., Nesterova, T., *et al.* (2008). Cohesins functionally associate with CTCF on mammalian chromosome arms. *Cell, 132*, 422-433.

Park, H., Li, Z., Yang, X.O., Chang, S.H., Nurieva, R., Wang, Y.H., Wang, Y., Hood, L., Zhu, Z., Tian, Q., *et al.* (2005). A distinct lineage of CD4 T cells regulates tissue inflammation by producing interleukin 17. *Nat. Immunol. 6*, 1133-1141.

Pearce, E.L., Mullen, A.C., Martins, G.A., Krawczyk, C.M., Hutchins, A.S., Zediak, V.P., Banica, M., DiCioccio, C.B., Gross, D.A., Mao, C.A., *et al.* (2003). Control of effector CD8+ T cell function by the transcription factor Eomesodermin. *Science, 302*, 1041-1043.

Peschon, J.J., Morrissey, P.J., Grabstein, K.H., Ramsdell, F.J., Maraskovsky, E., Gliniak, B.C., Park, L.S., Ziegler, S.F., Williams, D.E., Ware, C.B., *et al.* (1994). Early lymphocyte expansion is severely impaired in interleukin 7 receptor-deficient mice. *J. Exp. Med. 180*, 1955-1960.

Petrie, H.T., and Zuniga-Pflucker, J.C. (2007). Zoned out: functional mapping of stromal signaling microenvironments in the thymus. *Annu. Rev. Immunol. 25*, 649-679.

Phillips, J.E., and Corces, V.G. (2009). CTCF: master weaver of the genome. *Cell, 137*, 1194-1211.

Pui, J.C., Allman, D., Xu, L., DeRocco, S., Karnell, F.G., Bakkour, S., Lee, J.Y., Kadesch, T., Hardy, R.R., Aster, J.C., *et al.* (1999). Notch1 expression in early lymphopoiesis influences B versus T lineage determination. *Immunity, 11*, 299-308.

Radtke, F., Wilson, A., Stark, G., Bauer, M., van Meerwijk, J., MacDonald, H.R., and Aguet, M. (1999). Deficient T cell fate specification in mice with an induced inactivation of Notch1. *Immunity, 10*, 547-558.

Reinhardt, R.L., Liang, H.E., and Locksley, R.M. (2009). Cytokine-secreting follicular T cells shape the antibody repertoire. *Nat. Immunol. 10*, 385-393.

Ribeiro de Almeida, C., Heath, H., Krpic, S., Dingjan, G.M., van Hamburg, J.P., Bergen, I., van de Nobelen, S., Sleutels, F., Grosveld, F., Galjart, N., *et al.* (2009). Critical role for the transcription regulator CCCTC-binding factor in the control of Th2 cytokine expression. *J. Immunol. 182*, 999-1010.

Rodriguez, P., Bonte, E., Krijgsveld, J., Kolodziej, K.E., Guyot, B., Heck, A.J., Vyas, P., de Boer, E., Grosveld, F., and Strouboulis, J. (2005). GATA-1 forms distinct activating and repressive complexes in erythroid cells. *Embo J. 24*, 2354-2366.

Rothenberg, E.V., Moore, J.E., and Yui, M.A. (2008). Launching the T-cell-lineage developmental programme. *Nat. Rev. Immunol. 8*, 9-21.

Rothenberg, E.V., and Taghon, T. (2005). Molecular genetics of T cell development. *Annu. Rev. Immunol. 23*, 601-649.

Rubio, E.D., Reiss, D.J., Welcsh, P.L., Disteche, C.M., Filippova, G.N., Baliga, N.S., Aebersold, R., Ranish, J.A., and Krumm, A. (2008). CTCF physically links cohesin to chromatin. *Proc. Natl. Acad. Sci. U S A, 105*, 8309-8314.

Rutishauser, R.L., and Kaech, S.M. (2010) Generating diversity: transcriptional regulation of effector and memory CD8 T-cell differentiation. *Immunol. Rev. 235*, 219-233.

Saitoh, N., Bell, A.C., Recillas-Targa, F., West, A.G., Simpson, M., Pikaart, M., and Felsenfeld, G. (2000). Structural and functional conservation at the boundaries of the chicken beta-globin domain. *Embo J. 19*, 2315-2322.

Sandhu, K.S., Shi, C., Sjolinder, M., Zhao, Z., Gondor, A., Liu, L., Tiwari, V.K., Guibert, S., Emilsson, L., Imreh, M.P., *et al.* (2009). Nonallelic transvection of multiple imprinted loci is organized by the H19 imprinting control region during germline development. *Genes Dev. 23*, 2598-2603.

Schaerli, P., Willimann, K., Lang, A.B., Lipp, M., Loetscher, P., and Moser, B. (2000). CXC chemokine receptor 5 expression defines follicular homing T cells with B cell helper function. *J. Exp. Med. 192*, 1553-1562.

Schlissel, M.S. (2004). Regulation of activation and recombination of the murine Igkappa locus. *Immunol. Rev. 200*, 215-223.

Schmitt, T.M., Ciofani, M., Petrie, H.T., and Zuniga-Pflucker, J.C. (2004). Maintenance of T cell specification and differentiation requires recurrent notch receptor-ligand interactions. *J. Exp. Med. 200*, 469-479.

Schoenborn, J.R., Dorschner, M.O., Sekimata, M., Santer, D.M., Shnyreva, M., Fitzpatrick, D.R., Stamatoyonnapoulos, J.A., and Wilson, C.B. (2007). Comprehensive epigenetic profiling identifies multiple distal regulatory elements directing transcription of the gene encoding interferon-gamma. *Nat. Immunol. 8*, 732-742.

Sekimata, M., Perez-Melgosa, M., Miller, S.A., Weinmann, A.S., Sabo, P.J., Sandstrom, R., Dorschner, M.O., Stamatoyannopoulos, J.A., and Wilson, C.B. (2009). CCCTC-binding factor and the transcription factor T-bet orchestrate T helper 1 cell-specific structure and function at the interferon-gamma locus. *Immunity, 31*, 551-564.

Shimoda, K., van Deursen, J., Sangster, M.Y., Sarawar, S.R., Carson, R.T., Tripp, R.A., Chu, C., Quelle, F.W., Nosaka, T., Vignali, D.A., *et al.* (1996). Lack of IL-4-induced Th2 response and IgE class switching in mice with disrupted Stat6 gene. *Nature, 380*, 630-633.

Singer, A., Adoro, S., and Park, J.H. (2008). Lineage fate and intense debate: myths, models and mechanisms of CD4- versus CD8-lineage choice. *Nat. Rev. Immunol. 8*, 788-801.

Soler, E., Andrieu-Soler, C., de Boer, E., Bryne, J.C., Thongjuea, S., Stadhouders, R., Palstra, R.J., Stevens, M., Kockx, C., van Ijcken, W., *et al.* (2010) The genome-wide dynamics of the binding of Ldb1 complexes during erythroid differentiation. *Genes Dev. 24*, 277-289.

Song, S.H., Hou, C., and Dean, A. (2007). A positive role for NLI/Ldb1 in long-range beta-globin locus control region function. *Mol. Cell, 28*, 810-822.

Spilianakis, C.G., and Flavell, R.A. (2004). Long-range intrachromosomal interactions in thc T hclpcr typc 2 cytokine locus. *Nat. Immunol. 5*, 1017-1027.

Spilianakis, C.G., Lalioti, M.D., Town, T., Lee, G.R., and Flavell, R.A. (2005). Interchromosomal associations between alternatively expressed loci. *Nature, 435*, 637-645.

Splinter, E., Heath, H., Kooren, J., Palstra, R.J., Klous, P., Grosveld, F., Galjart, N., and de Laat, W. (2006). CTCF mediates long-range chromatin looping and local histone modification in the beta-globin locus. *Genes Dev. 20*, 2349-2354.

Staudt, V., Bothur, E., Klein, M., Lingnau, K., Reuter, S., Grebe, N., Gerlitzki, B., Hoffmann, M., Ulges, A., Taube, C., *et al.* (2010). Interferon-regulatory factor 4 is essential for the developmental program of T helper 9 cells. *Immunity, 33*, 192-202.

Stedman, W., Kang, H., Lin, S., Kissil, J.L., Bartolomei, M.S., and Lieberman, P.M. (2008). Cohesins localize with CTCF at the KSHV latency control region and at cellular c-myc and H19/Igf2 insulators. *Embo J. 27*, 654-666.

Szabo, P., Tang, S.H., Rentsendorj, A., Pfeifer, G.P., and Mann, J.R. (2000a). Maternal-specific footprints at putative CTCF sites in the H19 imprinting control region give evidence for insulator function. *Curr. Biol. 10*, 607-610.

Szabo, S.J., Kim, S.T., Costa, G.L., Zhang, X., Fathman, C.G., and Glimcher, L.H. (2000b). A novel transcription factor, T-bet, directs Th1 lineage commitment. *Cell. 100*, 655-669.

Szabo, S.J., Sullivan, B.M., Stemmann, C., Satoskar, A.R., Sleckman, B.P., and Glimcher, L.H. (2002). Distinct effects of T-bet in TH1 lineage commitment and IFN-gamma production in CD4 and CD8 T cells. *Science, 295*, 338-342.

Tabrizifard, S., Olaru, A., Plotkin, J., Fallahi-Sichani, M., Livak, F., and Petrie, H.T. (2004). Analysis of transcription factor expression during discrete stages of postnatal thymocyte differentiation. *J. Immunol. 173*, 1094-1102.

Taghon, T., Yui, M.A., Pant, R., Diamond, R.A., and Rothenberg, E.V. (2006). Developmental and molecular characterization of emerging beta- and gammadelta-selected pre-T cells in the adult mouse thymus. *Immunity, 24*, 53-64.

Taghon, T.N., David, E.S., Zuniga-Pflucker, J.C., and Rothenberg, E.V. (2005). Delayed, asynchronous, and reversible T-lineage specification induced by Notch/Delta signaling. *Genes Dev. 19*, 965-978.

Takeda, K., Tanaka, T., Shi, W., Matsumoto, M., Minami, M., Kashiwamura, S., Nakanishi, K., Yoshida, N., Kishimoto, T., and Akira, S. (1996). Essential role of Stat6 in IL-4 signalling. *Nature, 380*, 627-630.

Takemoto, N., Kamogawa, Y., Jun Lee, H., Kurata, H., Arai, K.I., O'Garra, A., Arai, N., and Miyatake, S. (2000). Cutting edge: chromatin remodeling at the IL-4/IL-13 intergenic regulatory region for Th2-specific cytokine gene cluster. *J. Immunol. 165*, 6687-6691.

Tanaka, S., Tsukada, J., Suzuki, W., Hayashi, K., Tanigaki, K., Tsuji, M., Inoue, H., Honjo, T., and Kubo, M. (2006). The interleukin-4 enhancer CNS-2 is regulated by Notch signals and controls initial expression in NKT cells and memory-type CD4 T cells. *Immunity, 24*, 689-701.

Thieu, V.T., Yu, Q., Chang, H.C., Yeh, N., Nguyen, E.T., Sehra, S., and Kaplan, M.H. (2008). Signal transducer and activator of transcription 4 is required for the transcription factor T-bet to promote T helper 1 cell-fate determination. *Immunity, 29*, 679-690.

Tolhuis, B., Palstra, R.J., Splinter, E., Grosveld, F., and de Laat, W. (2002). Looping and interaction between hypersensitive sites in the active beta-globin locus. *Mol. Cell, 10*, 1453-1465.

Tong, Y., Aune, T., and Boothby, M. (2005). T-bet antagonizes mSin3a recruitment and transactivates a fully methylated IFN-gamma promoter via a conserved T-box half-site. *Proc. Natl. Acad. Sci. U S A, 102*, 2034-2039.

Trifari, S., Kaplan, C.D., Tran, E.H., Crellin, N.K., and Spits, H. (2009). Identification of a human helper T cell population that has abundant production of interleukin 22 and is distinct from T(H)-17, T(H)1 and T(H)2 cells. *Nat. Immunol. s10*, 864-871.

Tsai, C.L., Rowntree, R.K., Cohen, D.E., and Lee, J.T. (2008). Higher order chromatin structure at the X-inactivation center via looping DNA. *Dev. Biol. 319*, 416-425.

Tydell, C.C., David-Fung, E.S., Moore, J.E., Rowen, L., Taghon, T., and Rothenberg, E.V. (2007). Molecular dissection of prethymic progenitor entry into the T lymphocyte developmental pathway. *J. Immunol. 179*, 421-438.

Usui, T., Nishikomori, R., Kitani, A., and Strober, W. (2003). GATA-3 suppresses Th1 development by downregulation of Stat4 and not through effects on IL-12Rbeta2 chain or T-bet. *Immunity, 18*, 415-428.

Usui, T., Preiss, J.C., Kanno, Y., Yao, Z.J., Bream, J.H., O'Shea, J.J., and Strober, W. (2006). T-bet regulates Th1 responses through essential effects on GATA-3 function rather than on IFNG gene acetylation and transcription. *J. Exp. Med. 203*, 755-766.

van Hamburg, J.P., de Bruijn, M.J., Ribeiro de Almeida, C., van Zwam, M., van Meurs, M., de Haas, E., Boon, L., Samsom, J.N., and Hendriks, R.W. (2008). Enforced expression of GATA3 allows differentiation of IL-17-producing cells, but constrains Th17-mediated pathology. *European journal of immunology, 38*, 2573-2586.

Veldhoen, M., Hocking, R.J., Atkins, C.J., Locksley, R.M., and Stockinger, B. (2006). TGFbeta in the context of an inflammatory cytokine milieu supports de novo differentiation of IL-17-producing T cells. *Immunity, 24*, 179-189.

Veldhoen, M., Uyttenhove, C., van Snick, J., Helmby, H., Westendorf, A., Buer, J., Martin, B., Wilhelm, C., and Stockinger, B. (2008). Transforming growth factor-beta 'reprograms' the differentiation of T helper 2 cells and promotes an interleukin 9-producing subset. *Nat. Immunol. 9*, 1341-1346.

Vogelzang, A., McGuire, H.M., Yu, D., Sprent, J., Mackay, C.R., and King, C. (2008). A fundamental role for interleukin-21 in the generation of T follicular helper cells. *Immunity, 29*, 127-137.

von Boehmer, H. (2005). Mechanisms of suppression by suppressor T cells. *Nat. Immunol. 6*, 338-344.

von Freeden-Jeffry, U., Vieira, P., Lucian, L.A., McNeil, T., Burdach, S.E., and Murray, R. (1995). Lymphopenia in interleukin (IL)-7 gene-deleted mice identifies IL-7 as a nonredundant cytokine. *J. Exp. Med. 181*, 1519-1526.

Wada, H., Masuda, K., Satoh, R., Kakugawa, K., Ikawa, T., Katsura, Y., and Kawamoto, H. (2008). Adult T-cell progenitors retain myeloid potential. *Nature, 452*, 768-772.

Wadman, I.A., Osada, H., Grutz, G.G., Agulnick, A.D., Westphal, H., Forster, A., and Rabbitts, T.H. (1997). The LIM-only protein Lmo2 is a bridging molecule assembling an erythroid, DNA-binding complex which includes the TAL1, E47, GATA-1 and Ldb1/NLI proteins. *Embo J. 16*, 3145-3157.

Wallace, J.A., and Felsenfeld, G. (2007). We gather together: insulators and genome organization. *Curr. Opin. Genet. Dev. 17*, 400-407.

Wendt, K.S., Yoshida, K., Itoh, T., Bando, M., Koch, B., Schirghuber, E., Tsutsumi, S., Nagae, G., Ishihara, K., Mishiro, T., *et al.* (2008). Cohesin mediates transcriptional insulation by CCCTC-binding factor. *Nature, 451*, 796-801.

Williams, A., Spilianakis, C.G., and Flavell, R.A. (2010). Interchromosomal association and gene regulation in trans. *Trends Genet. 26*, 188-197.

Wilson, C.B., Rowell, E., and Sekimata, M. (2009). Epigenetic control of T-helper-cell differentiation. *Nat. Rev. Immunol. 9*, 91-105.

Wurster, A.L., and Pazin, M.J. (2008). BRG1-mediated chromatin remodeling regulates differentiation and gene expression of T helper cells. *Mol. Cell Biol. 28*, 7274-7285.

Xie, X., Mikkelsen, T.S., Gnirke, A., Lindblad-Toh, K., Kellis, M., and Lander, E.S. (2007). Systematic discovery of regulatory motifs in conserved regions of the human genome, including thousands of CTCF insulator sites. *Proc. Natl. Acad. Sci. U S A, 104*, 7145-7150.

Xu, N., Donohoe, M.E., Silva, S.S., and Lee, J.T. (2007). Evidence that homologous X-chromosome pairing requires transcription and Ctcf protein. *Nat. Genet. 39*, 1390-1396.

Yamane, H., Zhu, J., and Paul, W.E. (2005). Independent roles for IL-2 and GATA-3 in stimulating naive CD4+ T cells to generate a Th2-inducing cytokine environment. *J. Exp. Med. 202*, 793-804.

Yamashita, M., Ukai-Tadenuma, M., Miyamoto, T., Sugaya, K., Hosokawa, H., Hasegawa, A., Kimura, M., Taniguchi, M., DeGregori, J., and Nakayama, T. (2004). Essential role of GATA3 for the maintenance of type 2 helper T (Th2) cytokine production and chromatin remodeling at the Th2 cytokine gene loci. *J. Biol. Chem. 279*, 26983-26990.

Yang, X.O., Panopoulos, A.D., Nurieva, R., Chang, S.H., Wang, D., Watowich, S.S., and Dong, C. (2007). STAT3 regulates cytokine-mediated generation of inflammatory helper T cells. *J. Biol. Chem. 282*, 9358-9363.

Yang, X.O., Pappu, B.P., Nurieva, R., Akimzhanov, A., Kang, H.S., Chung, Y., Ma, L., Shah, B., Panopoulos, A.D., Schluns, K.S., *et al.* (2008). T helper 17 lineage differentiation is programmed by orphan nuclear receptors ROR alpha and ROR gamma. *Immunity, 28*, 29-39.

Yu, W., Ginjala, V., Pant, V., Chernukhin, I., Whitehead, J., Docquier, F., Farrar, D., Tavoosidana, G., Mukhopadhyay, R., Kanduri, C., *et al.*

(2004). Poly(ADP-ribosyl)ation regulates CTCF-dependent chromatin insulation. *Nat. Genet. 36*, 1105-1110.

Yuan, J.S., Kousis, P.C., Suliman, S., Visan, I., and Guidos, C.J. Functions of notch signaling in the immune system: consensus and controversies. *Annu. Rev. Immunol. 28*, 343-365.

Yusufzai, T.M., Tagami, H., Nakatani, Y., and Felsenfeld, G. (2004). CTCF tethers an insulator to subnuclear sites, suggesting shared insulator mechanisms across species. *Mol. Cell, 13*, 291-298.

Zhang, D.H., Cohn, L., Ray, P., Bottomly, K., and Ray, A. (1997). Transcription factor GATA-3 is differentially expressed in murine Th1 and Th2 cells and controls Th2-specific expression of the interleukin-5 gene. *J. Biol. Chem. 272*, 21597-21603.

Zhang, D.H., Yang, L., Cohn, L., Parkyn, L., Homer, R., Ray, P., and Ray, A. (1999). Inhibition of allergic inflammation in a murine model of asthma by expression of a dominant-negative mutant of GATA-3. *Immunity, 11*, 473-482.

Zhang, F., and Boothby, M. (2006). T helper type 1-specific Brg1 recruitment and remodeling of nucleosomes positioned at the IFN-gamma promoter are Stat4 dependent. *J. Exp. Med. 203*, 1493-1505.

Zhao, H., Kim, A., Song, S.H., and Dean, A. (2006). Enhancer blocking by chicken beta-globin 5'-HS4: role of enhancer strength and insulator nucleosome depletion. *J. Biol. Chem. 281*, 30573-30580.

Zheng, W., and Flavell, R.A. (1997). The transcription factor GATA-3 is necessary and sufficient for Th2 cytokine gene expression in CD4 T cells. *Cell, 89*, 587-596.

Zhou, L., Ivanov, II, Spolski, R., Min, R., Shenderov, K., Egawa, T., Levy, D.E., Leonard, W.J., and Littman, D.R. (2007). IL-6 programs T(H)-17 cell differentiation by promoting sequential engagement of the IL-21 and IL-23 pathways. *Nat. Immunol. 8*, 967-974.

Zhu, J., Guo, L., Watson, C.J., Hu-Li, J., and Paul, W.E. (2001). Stat6 is necessary and sufficient for IL-4's role in Th2 differentiation and cell expansion. *J. Immunol. 166*, 7276-7281.

Zhu, J., Min, B., Hu-Li, J., Watson, C.J., Grinberg, A., Wang, Q., Killeen, N., Urban, J.F., Jr., Guo, L., and Paul, W.E. (2004). Conditional deletion of Gata3 shows its essential function in T(H)1-T(H)2 responses. *Nat. Immunol. 5*, 1157-1165.

Zhu, J., Yamane, H., and Paul, W.E. (2010). Differentiation of effector CD4 T cell populations. *Annu. Rev. Immunol. 28*, 445-489.

In: Zinc Fingers
Editors: R. Ciofani and L. Makrlik
ISBN 978-1-62100-230-7


*Chapter 2*

# Retroviral Zinc Knuckles in Eukaryotic Cellular Proteins

***Pablo Armas and Nora B. Calcaterra*** *

Instituto de Biología Molecular y Celular de Rosario (IBR),
Consejo Nacional de Investigaciones Científicas y Técnicas (CONICET) -
Área Biología General, Dpto. de Ciencias Biológicas,
Facultad de Ciencias Bioquímicas y Farmacéuticas,
Universidad Nacional de Rosario, Suipacha 531,
S2002LRK- Rosario, Argentina

## Abstract

Nucleocapsid (NC) proteins of retroviruses like HIV, derived from the proteolysis of the *gag*-encoded GAG polyprotein, are ≈ 50-amino acids long proteins that play pivotal roles in the viral lifecycle. These proteins contain one major functional domain consisting of two small zinc finger-like motifs named CCHC zinc knuckles conformed by the sequence Cys-$X_2$-Cys-$X_4$-His-$X_4$-Cys, where X can be any amino acid. The change of critical zinc binding amino acids; i.e., the three Cys or the His, to non-zinc binding residues results in virions that are defective in RNA packaging and replication. Interestingly, several proteins from eukaryotic cells contain from one to nine CCHC zinc knuckles and play

---

* E-mail: calcaterra@ibr.gov.ar

essential functions in the metabolism of nucleic acids. Although NC proteins were widely studied, scant information exists about such cellular proteins. Here we classify cellular CCHC zinc knuckle-containing proteins into two main groups. One group consists of proteins containing one to three CCHC zinc knuckles interspersed among other functional motifs. The other group is conformed by proteins that contain four or more tandem repeated CCHC zinc knuckles with relative short linkers between them, forming a multi-zinc knuckle core domain.

Moreover, we revise the function and manner of action of the most representative eukaryotic cellular CCHC-zinc knuckle proteins of each group, and carry out a structural and functional comparison between retroviral and cellular eukaryotic CCHC zinc knuckle-containing proteins. Finally, we point out some issues that should be explored to better understand the function of CCHC zinc knuckles in eukaryotic cellular proteins.

## Introduction

Zinc fingers are versatile and small structural protein domains that contain conserved cysteines and histidines coordinated to one or more zinc ions. Many proteins have multiple copies of zinc fingers; such proteins are commonly called zinc finger proteins. Zinc finger proteins perform a broad range of functions in various cellular processes, such as replication and repair, transcription and translation, metabolism and signaling, cell proliferation and apoptosis [1, 2].

Three major Zn finger classes can be distinguished according to the structural arrangement of cysteines and histidines zinc ligands and the role in living cells. The first class, CCHH finger, is by far the most abundant nucleic acid binding motif in human genome, and has the ability to bind to both DNA and RNA molecules. A second class, CCCH zinc finger, is present in proteins that regulate mRNA stability through AU-rich elements. The third class, CCHC zinc finger, is the main domain of retroviral nucleocapsid proteins (NCp) and binds to single-stranded nucleic acids.

## Retroviral Zinc Knuckles

Without exception, retroviral NCps (and their *gag* precursor proteins) from all strains of all known retroviruses contain one or two CCHC zinc fingers containing the same arrangement of Cys-$X_2$-Cys-$X_4$-His-$X_4$-Cys, in which X may be any amino acid residue. The CCHC motif contains about 18 amino acids (14 amino acids from the first Cys to the last Cys plus 2 surrounding residues at each side) and is thus shorter than CCHH- and CCCH-type domains (comprising about 30 residues).

The structure of the CCHC zinc finger is composed of two short β-strands connected by a turn (β-hairpin) followed by a short α-helix or loop [3]. The link between the second Cys and the His residue is just four amino acids long, much shorter than the classical DNA-binding zinc finger motif. This shortening precludes the formation of a α-helical nucleic acid binding "finger" and instead forms a 4-residue tight loop, which is referred in the literature as zinc "knuckle" domain [4]. Another name usually given to this structural domain is "*gag* knuckle" [1] since it contributes to the functionality of retroviral *gag*-encoded proteins.

# Zinc Knuckles in Eukaryotic Cellular Proteins

Several cellular zinc finger proteins presenting the motif $CX_2CX_4HX_4C$ have been identified in a wide range of eukaryotic organisms including trypanosomatids, yeast, worms, plants, and vertebrates, and have been shown to be involved in different cellular functions, most of them related with RNA metabolism.

Although the degree of conservation of the $CX_2CX_4HX_4C$ general arrangement is highly remarkable, the number of repetitions as well as their combination with other functional domains may vary among such cellular proteins.

To understand the structural and functional variety of cellular zinc-knuckle containing proteins we undertook a comprehensive survey and classified them in two main groups.

One group consists of proteins containing one to three CCHC zinc knuckles interspersed among other functional motifs. The other group is mainly conformed by proteins containing four or more tandem repeated CCHC knuckles separated by relative short amino acid sequences. An overview of proteins found is presented in Table 1.

## Table 1.

| Protein name | Gene name | Amino acid length | Number of CCHC | Sub-cellular location | Molecular Function | Additional functional motifs | Function | Organism | Database Reference |
|---|---|---|---|---|---|---|---|---|---|
| Pre-mRNA-splicing factor SLU7 | SLU7 | 586 | 1 | Nucleus, cytoplasm | Pre-mRNA 3'-splice site binding | NO | Pre-mRNA splicing, participates in the second catalytic step of pre-mRNA splicing, when the free hydroxyl group of exon I attacks the 3'-splice site to generate spliced mRNA and the excised lariat intron. | Homo sapiens | Uniprot O95391 |
| 5'-3' exoribonuclease 2 (DHM1-like protein) | XRN2 | 950 | 1 | Nucleus, nucleolus | Exonuclease | NO | 5'-3' exoribonuclease, may promote the termination of transcription by RNA polymerase II. | Homo sapiens | Uniprot Q9H0D6 |
| Splicing factor 1 (Mammalian branch point-binding protein) | SF1 | 638 | 1 | Nucleus | RNA binding | 1 KH domain<br>1 Pro rich region | Binds to the intron branch point sequence (BPS) 5'-UACUAAC-3' of the pre-mRNA. | Homo sapiens | Uniprot Q15637 |
| E3 ubiquitin-protein ligase | RBBP6 | 1792 | 1 | Nucleus, nucleolus | E3 ubiquitin-protein ligase | 1 DWNN domain.<br>1 RING-type zinc finger | Promotes the assembly of the p53/TP53-MDM2 complex, resulting in increase of MDM2-mediated ubiquitination and degradation of p53/TP53. | Homo sapiens | Uniprot Q7Z6E9 |

| Protein name | Gene name | Amino acid length | Number of CCHC | Sub-cellular location | Molecular Function | Additional functional motifs | Function | Organism | Database Reference |
|---|---|---|---|---|---|---|---|---|---|
| Xenopus-posterior | XPO | 555 | 1 | N/D | DNA binding | NO | Possible role in the organization of the anteroposterior axis during development. | Xenopus laevis | Uniprot P27536 |
| Zinc finger CCHC-type and RNA-binding motif 1 | ZCRB1 or MADP-1 | 217 | 1 | Nucleus, nucleoplasm | RNA binding | 1 RRM | Unknown. | Homo sapiens | Uniprot Q8TBF4 |
| Zinc finger CCHC domain containing 12 | ZCCHC12 | 402 | 1 | | Nucleic acid binding | NO | Transcriptional coactivator in the BMP-signaling pathway. | Homo sapiens | Uniprot Q6PEW1 |
| Nucleolar protein of 40 kDa | pNO40 or ZCCHC17 | 241 | 1 | Nucleus, nucleolus | RNA binding | 1 S1 motif | Ribosome maturation and/or biogenesis | Homo sapiens | Uniprot Q9NP64 |
| RNA-binding protein 4 | RBM4 | 364 | 1 | Nucleus, nucleolus, cytoplasm | RNA binding | 2 RRM | Aternative splice site selection during pre-mRNA processing. | Homo sapiens | Uniprot Q9BWF3 |
| Protein lin-28 homolog | LIN-28 | 209 | 2 | Cytoplasm, nucleus, nucleolus | DNA binding | 1 CSD | Translational enhancer, microRNA biogenesis suppressor. | Homo sapiens | Uniprot Q9H9Z2 |

**Table 1. (Continued)**

| Protein name | Gene name | Amino acid length | Number of CCHC | Sub-cellular location | Molecular Function | Additional functional motifs | Function | Organism | Database Reference |
|---|---|---|---|---|---|---|---|---|---|
| Branchpoint-bridging protein | BBP | 476 | 2 | Nucleus | Pre-mRNA branch point binding | 1 KH domain | Pre-mRNA retention and splicing | Saccharomyces cerevisiae | Uniprot Q12186 |
| Clipper | CLP | 296 | 2 | N/D | Endoribonuclease activity, nucleic acid binding | 5 CCCH zinc fingers | Pre-mRNA 3'-end processing | Drosophila melanogaster | Uniprot Q9VPT8 |
| Terminal uridylyltransferase 4 | TUT4 or ZCCHC11 | 1644 | 3 | Nucleus, cytoplasm. | RNA uridylyltransferase activity, nucleic acid binding | 2 PAP-associated domains 1 Ntr domain | Uridylyltransferase that acts as a suppressor of microRNA (miRNA) biogenesis by specifically mediating the terminal uridylation of some miRNAs. | Homo sapiens | Uniprot Q5TAX3 |
| Arginine methyltransferase-interacting RING finger protein 1 | **AIR1p** | 360 | 5 | Nucleus, cytoplasm | Nucleic acid binding, protein binding | NO | Component of the TRAMP complexes which have a poly(A) RNA polymerase activity and are involved in a post-transcriptional quality control mechanism limiting inappropriate expression of genetic information. | Sccharomyces cerevisiae | Uniprot P40507 |

| Protein name | Gene name | Amino acid length | Number of CCHC | Sub-cellular location | Molecular Function | Additional functional motifs | Function | Organism | Database Reference |
|---|---|---|---|---|---|---|---|---|---|
| Arginine methyltransferase-interacting RING finger protein 2 | **AIR2p** | 344 | 5 | Nucleus, cytoplasm | Nucleic acid binding, protein binding | NO | Component of the TRAMP complexes which have a poly(A) RNA polymerase activity and are involved in a post-transcriptional quality control mechanism limiting inappropriate expression of genetic information. | Sccharomyces cerevisiae | Uniprot Q12476 |
| Zinc finger CCHC domain-containing protein 9 | **ZCCHC9** | 271 | 4 | N/D | Nucleic acid binding | NO | Air1/2p homolog. | Homo sapiens | Uniprot Q8N567 |
| Germline RNA helicase 1 | GLH-1 | 763 | 4 | P-granule | RNA helicase | 1 helicase ATP-binding domain<br>1 DEAD box<br>1 helicase C-terminal domain | Translation regulation in P granules. | Caenoharbditis elegans | Uniprot P34689 |
| Germline RNA helicase 2 | GLH-2 | 974 | 6 | P-granule | RNA helicase | 1 helicase ATP-binding domain<br>1 DEAD box<br>1 helicase C-terminal domain | Translation regulation in P granules. | Caenoharbditis elegans | Uniprot Q966L9 |

**Table 1. (Continued)**

| Protein name | Gene name | Amino acid length | Number of CCHC | Sub-cellular location | Molecular Function | Additional functional motifs | Function | Organism | Database Reference |
|---|---|---|---|---|---|---|---|---|---|
| Germline RNA helicase 3 | GLH-3 | 720 | 2 | P-granule | RNA helicase | 1 helicase ATP-binding domain<br>1 DEAD box<br>1 helicase C-terminal domain | Translation regulation in P granules. | Caenorhabditis elegans | Uniprot O01836 |
| Germline RNA helicase 4 | GLH-4 | 1156 | 5 | P-granule | RNA helicase | 1 helicase ATP-binding domain<br>1 DEAD box<br>1 helicase C-terminal domain | Translation regulation in P granules. | Caenorhabditis elegans | Uniprot O76743 |
| Universal minicircle sequence-specific single-stranded DNA-binding protein | UMSBP | 116 | 5 | Mitocondria, Kinetoplast | Nucleic acid binding | NO | Kinetoplast DNA replication initiation and segregation. | Crithidia fasciculata | Uniprot O46363 |

| Protein name | Gene name | Amino acid length | Number of CCHC | Sub-cellular location | Molecular Function | Additional functional motifs | Function | Organism | Database Reference |
|---|---|---|---|---|---|---|---|---|---|
| Cellular nucleic acid binding protein. | **CNBP or ZNF9** | 177 | 7 | Nucleus, cytoplasm | ssDNA binding, RNA binding, nucleic acid chaperone | RGG box | Activates c-myc transcription by G4 formation promotion, regulates translation of ribosomal proteins mRNAs. | Homo sapiens | Uniprot P62633 |
| Zinc finger CCHC domain-containing protein 13 | ZCCHC13 or CNBP2 | 166 | 6 | N/D | Nucleic acid binding | RGG box | Unknown. | Homo sapiens | Uniprot Q8WW36 |
| CCHC-type zinc finger protein CG3800 | CG3800 | 165 | 6 | N/D | Nucleic acid binding | Gly-rich region | Unknown. | Dropsophila melanogaster | Uniprot Q8T8R1 |
| Glucose inhibition of gluconeogenic growth suppressor 2 protein | Gis2p | 153 | 7 | Cytoplasm | Single-stranded DNA and mRNA binding | NO | Translation regulation activity. | Sccharomyces cerevisiae | Uniprot P53849 |
| Cellular nucleic acid-binding protein homolog | **Byr3** | 179 | 7 | Nucleus | DNA binding | NO | Required for efficient yeast conjugation. | Schizosccharomyces pombe | Uniprot P36627 |
| Poly-zinc finger protein 1 | PZFP1 | 193 | 7 | Nucleus, cytoplasm | Nucleic acid binding | NO | Pre-mRNA processing | Trypanosoma cruzi | Uniprot O77233 |

**Table 1. (Continued)**

| Protein name | Gene name | Amino acid length | Number of CCHC | Sub-cellular location | Molecular Function | Additional functional motifs | Function | Organism | Database Reference |
|---|---|---|---|---|---|---|---|---|---|
| Cold shock domain protein 1 | CSDP1 | 299 | 7 | Nucleus, cytoplasm | RNA binding and chaperone | 1 CSD | RNA chaperone in response to cold stress, aiding in mRNA export from the nucleus to the cytoplasm and enhancing freezing tolerance. | Arabidopsis thaliana | NCBI Reference Sequence NP_195326 |
| Cold shock domain protein 2 | CSDP or GRP2 | 204 | 2 | Nucleus, cytoplasm | RNA binding and chaperone | 1 CSD | RNA chaperone in response to cold stress, aiding in mRNA export from the nucleus to the cytoplasm and enhancing freezing tolerance. | Arabidopsis thaliana | NCBI Reference Sequence NP_195580 |
| T. cruzi zinc finger protein 8 | TcZFP 8 | 192 | 8 | N/D | N/D | NO | Unknown. | Trypanosoma cruzi | [5] |
| DNA-binding protein HEXBP | **HEXBP** | 271 | 9 | Nucleus | DNA binding | NO | Binds to single-stranded DNA located in the conserved immediate 5'-UTR of the mRNA coding for the major surface glycoprotein GP63. | Leishmania mayor | Uniprot Q04832 |

*Group 1: proteins containing one to three CCHC zinc knuckles interspersed among other functional motifs.*

Many proteins have been described containing one to three dispersed CCHC zinc knuckles. The biological function of most of them is related with RNA metabolism. Following are summarized the most remarkable features and biological roles of representative proteins of this group.

## Proteins Containing a Unique CCHC Zinc Knuckle

*Pre-mRNA-splicing factor SLU7* (Slu7) is a predominantly nuclear protein. The CCHC domain spans from amino acids 118 to 135. This protein shuttles between the nucleus and the cytoplasm being this shuttling regulated by the CCHC zinc knuckle. Human Slu7 participates in the second catalytic step of pre-mRNA splicing, when the free hydroxyl group of exon 1 attacks the 3'-splice site to generate spliced mRNA and the excised lariat intron. It is required for holding exon 1 properly in the spliceosome and for correct AG identification when more than one possible AG exists in 3'-splicing site region [6]. Upon UV-C stress stimulus, the nuclear concentration of human SLU7 decreases affecting alternative splicing [7].

*5′-3′ exoribonuclease 2* (DHM1-like protein) is a ubiquitously distributed 950 aa-protein in which the CCHC zinc knuckle is conformed by amino acids 262 to 278 [8]. This protein is encoded by the XRN2 gene in humans, thus it is also named as Xrn2. DHM1-like protein locates to the nucleus and participates in the transcriptional termination of protein-encoding genes. During transcription termination, cleavage at the polyadenylation site liberates a 5' fragment which is subsequently processed to form the mature mRNA and a 3' fragment which remains attached to the elongating polymerase. The 3′ product of the cleavage is unstable and rapidly degraded by the exoribonucleolytic activity of DHM1-like protein, thus promoting termination of transcription [9].

*Splicing factor 1* (SF1) is alternatively named as mammalian branchpoint-binding protein, BBP or mBBP. Apart from a CCHC zinc knuckle (spanning residues 277 to 296), the mouse protein also contains a KH RNA binding domain. mBBP/SF1 locates to the nucleus in correspondence with the nuclear localization signal found in the N-terminal region. mBBP/SF1 also contains a C-terminal Pro-rich region. At least six isoforms were described as products of alternative splicing both in human and mouse. mBBP/SF1 functions at early stages of pre-mRNA splicing and contributes to splice site recognition by interacting with the essential splicing factor U2AF65 and binding to the intron branch site [10].

*E3 ubiquitin-protein ligase RBBP6,* also named as retinoblastoma binding protein 6 or p53-associated cellular protein of testis (PACT), is a nucleolar protein of 1790 amino acids that contains a CCHC zinc knuckle in position 160 to 177. It also contains an N-terminal DWNN domain and a RING-type zinc finger motif. RBBP6 interacts with retinoblastoma 1 (RB1) and p53 [11]. RBBP6 plays a role as a scaffold protein to promote the assembly of the p53/TP53-MDM2 complex, resulting in increase of MDM2-mediated ubiquitination and degradation of p53/TP53. It may function as negative regulator of p53/TP53, leading to both apoptosis and cell growth retardation [12].

*Xenopus-posterior* (Xpo) is a gene activated during *Xenopus* embryo development at or shortly after the midblastula transition (MBT). Its expression is asymmetrically localized in the posterior ectoderm and mesoderm after gastrulation. Xpo has a predicted size of 555 amino acid residues with a CCHC zinc knuckle located in position 414 to 429, near the C-terminus. Xpo expression is induced in ectodermal explants by treatment with basic fibroblast growth factor (bFGF) suggesting a possible role in the organization of the anteroposterior axis during development. [13]. Besides, Xpo was found to be induced by ectopic expression of functional Xwnt-8 in the dorsal marginal zone of the embryo during gastrulation [14].

*Zinc finger CCHC-type and RNA-binding motif 1* (ZCRB1 or MADP-1) was cloned while isolating human morphine-dependence-related genes. ZCRB1 encodes a 217-residue protein with an RNA recognition motif (RRM) at the N-terminus and a CCHC zinc finger motif spanning positions 106 to 124. ZCRB1 is conserved from zebrafish to human and shares homology with cold-inducible RNA-binding protein. ZCRB1 is located in the nucleoplasm, but outside the nucleolus. ZCRB1 gene expression was stimulated by morphine, inhibited at low temperatures, and up-regulated at high temperatures in SH-SY5Y neural cells [15].

*Zinc finger CCHC domain containing 12* (ZCCHC12) has been identified in mouse as a transcriptional co-activator of bone morphogenetic protein (BMP) signaling, and human ZCCHC12 was reported to be related to non-syndromic X-linked mental retardation. ZCCHC12 is a 402-amino acid protein with a CCHC zinc knuckle at position 345 – 362. It contains a nuclear localization signal in the middle of its sequence. ZCCHC12 is highly expressed in human brain and is specifically expressed in neuroepithelium of forebrain, midbrain, and diencephalon regions of mouse E10.5 embryos.

ZCCHC12 enhances the transcriptional activities of activator protein 1 (AP-1) and cAMP response element binding protein (CREB) as a coactivator [16].

*Nucleolar protein of 40 kDa* (pNO40) is a human nucleolar protein identified through yeast two hybrid interaction screen of a human kidney cDNA library using pinin (pnn) protein as the bait. pNO40 is a protein of 241 amino acids conserved from zebrafish to human. It contains several domains including a ribosomal protein S1 RNA binding region at the N-terminus, a CCHC type zinc finger spanning amino acids 133 to 146 in the middle of the protein, and two clusters of basic amino acids at the C-terminus that seem to function as bipartite nuclear and nucleolar targeting sequences. pNO40 localized to nucleoli of diverse cultured cells, with some concentration in the granular component of nucleoli, relating its function to ribosome maturation and/or biogenesis [17].

*RNA-binding protein 4* (RBM4), also known as Lark, was described initially as having a role in circadian rhythm control in *Drosophila*. Human RBM4 contains two RNA recognition motifs (RRM) at the N-terminus and a CCHC zinc knuckle in the middle of the protein (amino acids 160 to 176). RBM4 is expressed throughout the nucleoplasm, but is concentrated in nucleoli and in speckles wherein co-localizes with SR-proteins. Likewise many RNA processing proteins, RBM4 has a dynamic subcellular expression pattern and it was observed also in the cytoplasm. Data from mammalian cells revealed that RBM4 is an RNA-binding protein involved in diverse cellular processes that include alternative splicing of pre-mRNA, translation, and RNA silencing [18].

## Proteins Containing Two or Three CCHC Zinc Knuckles

The variety of eukaryotic cellular proteins containing two or three CCHC zinc knuckles is much smaller than that of proteins with a single one. Those for which there exists available information are briefly described below.

*Protein lin-28 homolog* (lin-28) is a protein homolog to the heterochronic gene lin-28 product found in the nematode *Caenorhabditis elegans* as a regulator of developmental timing. Lin-28 homologues are expressed in diverse animals, including *Drosophila*, *Xenopus*, mouse, zebrafish and human. Lin-28 homologues are proteins of about 200 amino acids (209 in the human protein) characterized by the presence of a cold shock domain (CSD) at the N-terminal region and two CCHC zinc knuckles in tandem at the C-terminal region (spanning positions 137 to 176 in the human protein). Lin-28 homolog protein was reported as a translational enhancer, driving specific mRNAs to polysomes and increasing the number of initiation events per molecule of mRNA thus increasing the efficiency of protein synthesis. It may also act as a suppressor of microRNA (miRNA) biogenesis by specifically binding the precursor let-7 (pre-let-7) and recruiting ZCCHC11/TUT4 uridylyltransferase (see below), leading to the terminal uridylation of pre-let-7. Uridylated pre-let-7 miRNAs fail to be processed by Dicer and undergo degradation [19].

*Branchpoint-bridging protein* (BBP) is a protein found in *Saccharomyces cerevisiae* that is also named splicing factor 1 (ScSF1). BBP/ScSF1 is 476 amino acid long (shorter than the ortholog found in mammals) and contains a KH RNA binding domain near the N-terminus, two central CCHC zinc knuckles in tandem (amino acids 271 to 314) and a C-terminal proline-rich region. BBP/ScSF1 locates to the nucleus and participates both in pre-mRNA retention and splicing [20].

*Clipper* (CLP) is a *Drosophila melanogaster* gene product homolog to a component of the multimeric protein complex known as cleavage and polyadenylation specificity factor (CPSF), an essential component of the mammalian pre-mRNA 3'-end processing machinery. CLP is a 296-amino acids long protein that possesses five N-terminal CCCH zinc fingers with endoribonucleolytic activity and two C-terminal CCHC zinc knuckles (amino acids 191 to 204 and 268 to 281), conferring a binding preference for RNAs that contain G- and/or C-rich clusters [21].

*Zinc finger CCHC domain-containing protein 11* (ZCCHC11) is a 3′ terminal uridylyl transferase (TUTase 4) responsible for uridylation and blockade of pre-let-7 miRNA maturation in embryonic stem (ES) cells and certain cancers mediated by Lin28 RNA-binding proteins and proto-oncogenes. ZCCHC11 activity is UTP dependent, selective for let-7, and recruited by Lin28. Knockdown of either Zcchc11 or Lin28, or overexpression of a catalytically inactive TUTase, relieves the selective inhibition of let-7 processing and leads to the accumulation of mature let-7 miRNAs and repression of let-7 target genes. Zcchc11 is a protein of 1644 amino acids predicted to contain a CCHH zinc finger near its N-terminus, two poly(A) polymerase (PAP) domains, a nucleotidyltransferase (Ntr) domain, and three CCHC zinc knuckles interspersed among PAP and Ntr domains [22]. ZCCHC11 was also suggested to fine tune cytokine expression (interleukine 6 or IL-6) by uridylating miRNA-26a that targets IL-6 [23].

*Group 2: proteins that contain four or more tandem repeated CCHC zinc knuckles*

This group is represented by proteins that contain four or more tandem repeated CCHC zinc knuckles with relative short linkers between them. The multi-zinc knuckle core is predominant or exclusive in the proteins' length and plays a central role in protein molecular functions that are mainly related with nucleic acids metabolism. The most representative and best described examples are detailed below. Proteins are ordered according to their increasing number of CCHC zinc knuckles. In some cases, two or more proteins containing different number of zinc knuckles belonging to a same family are described together.

*Arginine methyltransferase-interacting RING finger proteins 1 and 2* (Air1p/Air2p) are two proteins found in *Saccharomyces cerevisiae* that interact with Hmt1p major arginine methyltransferase and inhibit methylation of the Arg-Gly-Gly rich motif (RGG box) of the heterogeneous nuclear ribonucleoprotein (hnRNP) Npl3p, homologous to human hnRNP A1. Air1p and Air2p may affect mRNA transport out from the nucleus by regulating the arginine methylation state of hnRNPs [24]. Air1p and Air2p are proteins of 360 and 344-amino acids, respectively, and were originally described to contain a RING finger domain near the N- terminus, but later it was found to contain five tandem repeats of CCHC zinc knuckles [25]. These proteins have been described forming part of the Trf4/5p–Air1/2p–Mtr4p polyadenylation (TRAMP 4/5) complex, together with the poly(A) polymerases Trf4/5p and the putative RNA helicase Mtr4p [25-27]. The TRAMP complexes are considered to be cofactors that stimulate the nuclear exosome. Air1p or Air2p may recruit the TRAMP complex to target RNAs via RNA interactions through the most N-terminal three CCHC zinc knuckles, allowing the poly(A) polymerase Trf4p, associated to Air1/2p through the most C-terminal two CCHC zinc knuckles, to add a single-stranded poly(A) tail, making the RNA a better substrate for 3' degradation by the exosome [26, 27].

*Zinc finger CCHC domain-containing protein 9* (ZCCHC9) is a 271-amino acid long human nucleolar protein expressed in brain that contains four CCHC zinc knuckles with sequence homology to yeast Air1p/Air2p [28]. Mouse ZCCHC9 gene shows the highest expression levels in brain, particularly in the cortex, and in testicles [29].

*Germline RNA helicases* (GLHs) are a family of proteins identified in *Caenorhabditis elegans* that contain an N-terminal region rich in FGG repeats, two to six tandem repeated CCHC zinc knuckles, one helicase ATP-binding domain, one DEAD box, and one helicase C-terminal domain [30]. The different members of the family, GLH-1, 2, 3 and 4, are conformed by the same protein motifs but contain variable lengths of N-terminal FGG-rich domains and four, six, two, and five CCHC zinc knuckles, respectively. In GLH-1 and 2, FGG repeats are interspersed with CCHC zinc knuckles [31]. All four GLH proteins associate with *C. elegans* germ granules or P granules, but GLH-1 is the key member of the family because *glh-1* loss-of-function causes temperature-dependent sterility, which is due to reduced germ-cell proliferation and impaired formation of both sperm and oocytes [32]. The other *glh* genes are not essential, although GLH-4 seems to play redundant role with GLH-1 since loss-of-function of both genes causes temperature independent sterility [33]. Each GLH may bind different sets of RNAs and regulate their translation at designated times during germline development [30].

*Universal minicircle sequence-specific single-stranded DNA-binding protein* (UMSBP) is a 116-amino acid ling protein found in the trypanosomatid *Crithidia fasciculata* that binds to the conserved universal minicircle sequence (UMS) wherein the kinetoplast DNA (kDNA) minicircle light chain initiates replication [34, 35] and, because of this, it has been implicated in kDNA replication initiation and segregation [36]. UMSBP consists exclusively of five tandem repeats of CCHC zinc knuckles joint by short linker peptides. The redox state of the Cys residues seems to be a cellular mechanism to control UMSBP activity since reduced protein is monomeric and active while oxidized protein is dimeric and inactive [37, 38]. Probable homologues of this protein are PDZ5 in *Trypanosoma cruzi* [39] and TbZFP protein in *Trypanosoma brucei* [40].

*Cellular nucleic acid binding protein* (CNBP or ZNF9) has been originally found as a human single-stranded nucleic acid binding protein of 177 amino acids isolated from an expression library screening with a promoter fragment containing a sterol regulatory element from the hydroxymethylglutaryl-coenzyme A reductase gene [41]. Later, CNBP has been implicated in vertebrate craniofacial development [42-44] and in myotonic dystrophy type 2 (DM2) [45] and sporadic inclusion body myositis (sIBM) [46] human diseases. Proteins homologous to CNBP have been found in chordates, multicellular and unicellular non-vertebrates, but not in plants or bacteria , and it has also received the name of zinc finger protein 9 (ZNF9) [47]. CNBP protein family shows a high degree of identity among vertebrates (identities $\geq$ 65%), and lower values among other eukaryotic sequences (identities $\leq$ 35% for arthropod insects and simple unicellular eukaryotes) [47]. CNBP has a strikingly conserved modular structure with seven tandem CCHC zinc knuckle repeats joint by short (4 to 10 amino acids) peptide linkers, with the exception of the linker between the first and second CCHC repeats, that are separated by a region of 17 to 35 amino acids that contains a glycine/arginine-rich region highly similar to the RGG box of other RNA-binding proteins. In addition, CNBP contains a putative nuclear localization signal in the short linker joining the third and fourth CCHC zinc knuckles, what is consistent with its nuclear/cytoplasmic localization, a putative PEST (Pro, Glu, Ser or Thr-enriched) proteolysis region adjacent to the RGG box in the linker joining the first and second CCHC zinc knuckles, and a conserved cAMP-dependent protein kinase (PKA) phosphorylation site immediately to the C-terminal of the last CCHC zinc knuckle [47]. CNBP has been implicated in the transcriptional and posttranscriptional control of many genes through binding to a great variety of single-stranded nucleic acids target sequences, ranging from DNA regulatory sites in gene promoters to RNA translational regulatory elements in untranslated regions of mRNAs. It has been reported that the preferred CNBP targets are single-stranded nucleic acids containing G-rich unpaired stretches constrained by an organized sequence environment, promoting rearrangements of nucleic acid secondary structure in an ATP-independent manner, acting as a nucleic acid chaperone [48]. The human *c-myc* proto-oncogene, has been proposed to be controlled by the binding of CNBP to a G-rich sequence present in the promoter and catalyzing the formation of intramolecular four-stranded G-quadruplexes or G4s [49]. The CCHC zinc knuckles are required but not individually essential and contribute to achieving maximal nucleic acid binding and chaperone activities. However, the N-terminal zinc knuckle and the RGG box are essential for RNA binding and nucleic acid chaperone activity for catalyzing G4-folding [49, 50]. It has been proposed that the CCHC and RGG motifs form multiple pockets that interact with unpaired nucleobases (mainly guanine) and cause the co-folding of the protein with a G-rich single-stranded nucleic acid. CNBP biological function in neurocranial development, DM2 and sIBM has been integrated and related to cell death and proliferation rates [47]. Low levels of CNBP may reduce the rate of global protein synthesis by reduction in the translation of components of the translation machinery, thereby reducing proliferation and increasing apoptosis. Conversely, CNBP might affect transcription of genes required for cell proliferation.

*Zinc finger CCHC domain-containing protein 13* (ZCCHC13 or CNBP2) is a 166-amino acid long human protein structurally homolog to CNBP also known as CNBP2. This protein contains only four central complete CCHC zinc knuckles, being the two most N-terminal ones degenerate (pseudo-CCHC) due to amino acid replacements in the conserved Cys and His positions (the first Cys is replaced by Phe in the first knuckle and His is replaced by Arg in the second one); moreover, ZCCHC13 lacks the most C-terminal CCHC motif. ZCCHC13 has no described function and has only been identified in a comparative analysis of the human and the chimpanzee genome sequences as one of nine human-specific frameshift mutations [51]. ZCCHC13 is mapped in the X-inactivation region in mouse and human [52]. The chimpanzee ZCCHC13 has the same CCHC and pseudo-CCHC zinc knuckles as the human protein and possesses the most C-terminal CCHC zinc knuckle, which is lost in the human version by an insertion of cytosine nucleotide between codons 150 and 151 (Hahn and Lee. 2005. Bioinformatics. 21(1): i186-i194). The mouse ZCCHC13 has seven complete CCHC zinc knuckles and is mainly expressed in adult mouse testis [52]. All human and mouse expressed sequence tags with defined tissue source were isolated from testis, suggesting an involvement in the reproductive process [51]. However, a recent report indicate that ZCCHC13 is one of the proteins differentially expressed in glioma vessels [53].

*CCHC-type zinc finger protein CG3800* is a protein from *Drosophila melanogaster* of unknown function that was predicted to have homology with CNBP based on sequence similarity. CG3800 encodes a polypeptide of 165 amino acids showing a modular organization highly similar to CNBP. However, CG3800 lacks one CCHC zinc knuckle and the domain between the first and second CCHC zinc knuckles is mainly conformed by Gly residues.

*Glucose inhibition of gluconeogenic growth suppressor 2 protein* (Gis2p) is the product of a *Saccharomyces cerevisiae* gene that has been described as the CNBP ortholog [54]. Gis2p target mRNAs contain stretches of G(A/U)(A/U) trinucleotide repeats located in coding sequences. Using this sequence target motif, the CNBP predicted targets belong to the same functional categories as those for recognized by Gis2p in yeast, indicating functional conservation. It was suggested that CNBP and Gis2p may coordinate the expression of conserved RNA regulons, primarily by reducing mRNA and protein levels of genes required for ribosome assembly and by selectively up-regulating protein levels of myosin [54]. Accordingly, as it was previously reported for CNBP, Gis2p interacts with the protein translation machinery, and its expression in HEK293T cells activates cap-independent translation driven by the 5'UTR of the ODC gene [55].

*Byr3* is a *Schizosaccharomyces pombe* gene found as a multicopy suppressor of the sporulation defects of *rasl* null diploids. It encodes a 179-amino acid protein homolog to CNBP with seven tandem repeated CCHC zinc knuckles. Byr3 lacks the RGG box between the first and second zinc knuckles but presents a longer linker between the fourth and fifth CCHC zinc knuckles with a small RGG box-like motif; moreover, the linker between the first and second CCHC zinc knuckles is shorter than the one present in CNBP. Later it was shown that *Byr3* is not an essential gene for sporulation, but its product appears to be required for efficient yeast conjugation, being CNBP capable of restoring conjugation of *byr3* null strain towards wild-type levels [56]. A recent report indicates that *Byr3* protein interacts with the ribonuclease III Dicer (Dcr1) linking this enzyme, required for chromosome segregation and gene silencing, to the transcriptional and translational machineries [57].

*Poly-zinc finger protein 1* (PZFP1) is a protein found in *Trypanosoma cruzi* that has 191 amino acids and contains seven tandem repeated CCHC zinc knuckles [58]. Compared to CNBP, PZFP1 has a shorter linker peptide between zinc knuckles 1 and 2, no RGG box, but interspersed RGG-like trinucleotides in most of the zinc knuckles linkers. PZFP1 specifically interacts with single-stranded DNA or RNA oligonucleotides carrying the sequence 5'-GGGGT$(G/T)_2$G$(G/T)_2$-3'. The protein was localized mainly in the cytoplasm and nucleus, but not in the kinetoplast, precluding a role in minicircles replication. PZFP1 interacts specifically with a *T. cruzi* serine–arginine-rich protein (TcSR), suggesting a role in pre-mRNA processing [58].

*Cold shock domain protein 1* (CSDP1) is a 299 amino acids protein of *Arabidopsis thaliana* that contains a cold shock domain (CSD) in the N-terminal half and seven tandem repeated CCHC zinc knuckles interspersed with Gly-rich sequences in the C-terminal half. CSDP1 functions as an RNA chaperone in the response of *Arabidopsis* to cold stress, aiding in mRNA export from the nucleus to the cytoplasm and enhancing freezing tolerance. CSDP2, another *Arabidopsis* CSD protein, has a similar structure but is shorter (204 amino acids) and only possesses two CCHC zinc knuckles in the Gly-rich C-terminal half [59]. CSDP2 has much lower RNA binding and chaperone activities as a consequence of the reduced number of CCHC zinc knuckles, directly relating the number and modular arrangement of CCHC zinc knuckles to the RNA binding and chaperone activity [60].

*T. cruzi zinc finger protein 8* (TcZFP8) is a *Trypanosoma cruzi* 192 amino acids nuclear protein containing eight tandem repeated CCHC zinc knuckles with no other predicted protein domain. TcZFP8 is encoded in the same region of the XX chromosome of *T. cruzi* as PDZ5 and PZFP1 previously described [5].

*Hexamer binding protein* (HEXBP) is a protein from the protozoan parasite *Leishmania major* capable of binding to the hexanucleotide tandem repeats present within the conserved immediate 5'-UTR of the mRNA coding for the major surface glycoprotein GP63 [61]. HEXBP is a 271 amino acids protein that presents nine tandem repeated CCHC zinc knuckles with linkers longer than those present in CNBP (12 to 29 amino acids). HEXBP function is unknown, since it has no effect on GP63 expression, nor is essential for *L. major* growth [61].

## Retroviral *vs.* Eukaryotic Zinc Knuckles

Retroviral NCp and CNBP CCHC zinc knuckles share arrangements of amino acids with the general sequence: "–Cys–ϕ–X–Cys–Gly–±–X–Gly–His–$X_3$–δ–Cys–", where X is a variable amino acid, ϕ is an aromatic residue, ± is a charged amino acid, and δ is a carbonyl-containing residue [62]. An exhaustive analysis of the CCHC zinc knuckle sequences from the eukaryotic cellular proteins described above rendered the novel consensus sequence pattern shown in Figure 1.

A sequence logo was created to determine not only the consensus sequence but also subtle sequence patterns obtained from the relative frequency of amino acids at every position. Apart from a remarkable conservation of the cysteines (positions 3, 6, and 16) and histidine (position 11), conserved glycine within the second loop at positions 7 and 10 were found. The presence of these glycine residues may reflect a requirement for small and sterically nondemanding residues at these positions of the compact motif structure, as it was suggested for the retroviral motifs [64]. An aromatic residue is frequently found at position 4, hydrophobic residues are found less frequently at position 12, and serine or alanine are frequently found at position 13. These residues form a hydrophobic cleft in the HIV-1 Gag motif in which specific hydrogen bonds with a guanine of the nucleic acid ligand [64, 65]. Gag motifs contain a basic residue immediately before the first cysteine that specifically contacts with guanine of ligands. This residue is not conserved in the CCHC motifs of cellular proteins; instead, basic residues that might serve the same function are frequently found at position 14 of eukaryotic cellular CCHC motifs. Acidic residues are frequently found at position 15; however, their roles have not yet elucidated. Position 17 shows the presence of a proline residue that may reflect a structural conservation of a turn in the backbone of the motif. Finally, positions 5 and 8 show the presence of polar or charged residues able to form hydrogen bonds with nitrogen and carbonyl groups of ligand nucleobases, as it was reported for retroviral NCp [66, 67]. Altogether, the new general CCHC consensus sequence is: "–X–X–Cys–ϕ–ψ–Cys–Gly–

ψ–X–Gly–His–ω–(A/S)–(+)–(–)–Cys–Pro–X–", where X is a variable amino acid, ϕ an aromatic residue, ψ a polar or charged residue able to form hydrogen bonds, ω a hydrophobic residue, (A/S) alanine or serine, (+) a residue with positive charge, and (–) a residue with negative charge. The new consensus shows that 9 out of the 14 positions of the CCHC motifs core are conserved between retroviral NCp and zinc knuckle-containing eukaryotic cellular proteins.

The remarkable conservation strongly suggests highly conserved biochemical and structural features, as well as biological functions. This notion is supported by the finding that most of the CCHC zinc knuckles from human CNBP (six of seven) are capable of replacing human immunodeficiency virus type 1 (HIV-1) NC without affecting viral structure and function [62]. HIV-1 NC protein binds its target in a mutually induced fashion, in which NC protein pulls unstacked guanosines into hydrophobic clefts on the surface of the zinc knuckles [66-68]. Thus, cellular proteins may bind to its targets involving the entrance of guanosine nucleobases to the pockets formed by the CCHC zinc knuckles. The concept that CCHC zinc knuckle-containing proteins bind to single-stranded nucleic acids showing specific arrays of guanosines has enabled to use the CCHC zinc knuckles as modules to rationally design specific single-stranded nucleic acid binding peptides [69]). As regarding CCHC cellular functions, proteins containing CCHC zinc knuckles have nucleic acid chaperone activity [59, 60], and this activity is directly related to the number of zinc knuckle repeats. NCp has potent nucleic acid chaperone activity, which is central for its functionality and dependent on zinc knuckles [70]. The HIV-1 NC nucleic acid chaperone activity is based on two related activities: hybridization and helix destabilization. The first is sequence-unspecific and depends on the positive charges of basic amino acids surrounding the CCHC zinc knuckles. Helix destabilization activity depends on the zinc knuckles because of their preference for binding single-stranded (unpaired) nucleotides, mainly guanine.

In cellular proteins, auxiliary basic amino acids (mainly Arg)-enrich motifs may act in a sequence-unspecific way while CCHC zinc knuckles contribute in a sequence-specific fashion to the nucleic acid chaperone activity by binding to and stabilizing single-stranded unpaired nucleobases [48, 50]. Throughout its nucleic acid chaperone activity, NCp perturbs the equilibrium of intramolecular G4 formation [71, 72].

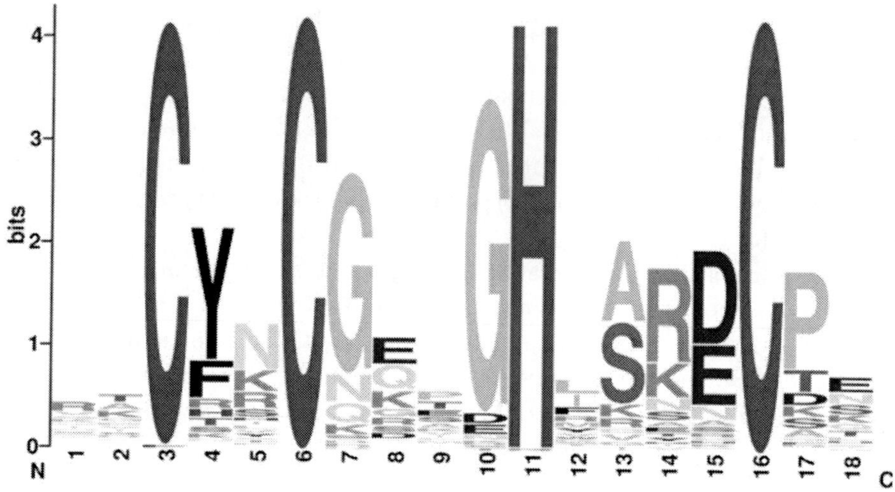


$$X-X-C-\phi-\psi-C-G-\psi-X-G-H-\omega-{}^{A}\!/_{S}-(+)-(-)-C-P-X$$

Figure 1. Consensus sequence pattern for eukaryotic CCHC zinc knuckle. The sequences of each individual CCHC zinc knuckle from the cellular proteins described in this chapter were aligned considering 18 amino acids motifs consisting of the 14 core residues from the first to the last conserved cysteines plus 2 surrounding residues at each side. The alignment was used to run the online software WebLogo 3 (http://weblogo.berkeley.edu/logo.cgi) [63], obtaining the sequence logo shown in the figure. Below the logo is the deduced general consensus sequence where C is a cysteine, H is a histidine, G is a glycine, P is a proline, X is a variable amino acid, $\phi$ is an aromatic residue, $\psi$ is a polar or charged residue able to form hydrogen bonds , $\omega$ is a hydrophobic residue, (A/S) is alanine or serine, (+) is a basic residue with positive charge, and (–) is an acidic residue with negative charge.

As we have recently described, CNBP promotes the formation of intramolecular G4 through its nucleic acid chaperone activity [49], thus suggesting that other zinc knuckle-containing eukaryotic cellular proteins may modulate G4 formation through the preference guanine nucleobases binding.

## Conclusion

In this chapter, an analysis of the main features of zinc knuckle-containing proteins identified in eukaryotic cells was conducted on the base on main biological function descriptions and zinc-knuckle motif structural comparisons. The degree of conservation found among zinc knuckle amino acid sequences conforming proteins of highly divergent evolutionary origins is outstanding. This finding strongly suggests that CCHC zinc knuckles may be considered as a unit of molecular evolution useful for the study of protein evolution and for large-scale annotation of genome sequences.

Fine structural information about CCHC zinc knuckles in retroviral NCp has led to a quite complete understanding of the structure-function relationship of this protein. Although, information on cellular zinc knuckle-containing proteins has increased significantly in recent years, most studies have focused on the functional features of these proteins and, especially for multi-zinc knuckle proteins, detailed structural or biochemical studies has not been carried out yet. This lack of information makes difficult to fully understand the structure-function relationship of this kind of proteins.

The only attempts to study structural features of multi-zinc knuckle proteins have been made very recently for the yeast Air2p through crystallographic studies [73], and for the human homolog ZCCHC9 through nuclear magnetic resonance [28]. Therefore, more structural data of multi-zinc knuckle proteins, free and bound to their targets, are imperative in order to better understand the way of action of this protein domain in cellular proteins.

# References

[1] Krishna, S. S., Majumdar, I., and Grishin, N. V. (2003). Structural classification of zinc fingers: survey and summary. *Nucleic Acids Res.*, 31, 532-550.

[2] Laity, J. H., Lee, B. M., and Wright, P. E. (2001). Zinc finger proteins: new insights into structural and functional diversity. *Curr.Opin.Struct.Biol.*, 11, 39-46.

[3] Summers, M. F., Henderson, L. E., Chance, M. R., Bess, J. W., Jr., South, T. L., Blake, P. R., Sagi, I., Perez-Alvarado, G., Sowder, R. C., III, Hare, D. R., and . (1992). Nucleocapsid zinc fingers detected in retroviruses: EXAFS studies of intact viruses and the solution-state structure of the nucleocapsid protein from HIV-1. *Protein Sci.*, 1, 563-574.

[4] South, T. L., Kim, B., Hare, D. R., and Summers, M. F. (1990). Zinc fingers and molecular recognition. Structure and nucleic acid binding studi.

[5] Ericsson, A. O., Faria, L. O., Cruz, W. B., Martins, d. S., and Lima, B. D. (2006). TcZFP8, a novel member of the Trypanosoma cruzi CCHC zinc finger protein family with nuclear localization. *Genet.Mol.Res.*, 5, 553-563.

[6] Chua, K. and Reed, R. (1999). The RNA splicing factor hSlu7 is required for correct 3' splice-site choice. *Nature*, 402, 207-210.

[7] Shomron, N., Alberstein, M., Reznik, M., and Ast, G. (2005). Stress alters the subcellular distribution of hSlu7 and thus modulates alternative splicing. *J.Cell Sci.*, 118, 1151-1159.

[8] Zhang, M., Yu, L., Xin, Y., Hu, P., Fu, Q., Yu, C., and Zhao, S. (1999). Cloning and mapping of the XRN2 gene to human chromosome 20p11.1-p11.2. *Genomics*, 59, 252-254.

[9] Gromak, N., West, S., and Proudfoot, N. J. (2006). Pause sites promote transcriptional termination of mammalian RNA polymerase II. *Mol.Cell Biol.*, 26, 3986-3996.

[10] Wang, X., Bruderer, S., Rafi, Z., Xue, J., Milburn, P. J., Kramer, A., and Robinson, P. J. (1999). Phosphorylation of splicing factor SF1 on Ser20 by cGMP-dependent protein kinase regulates spliceosome assembly. *EMBO J.*, 18, 4549-4559.

[11] Simons, A., Melamed-Bessudo, C., Wolkowicz, R., Sperling, J., Sperling, R., Eisenbach, L., and Rotter, V. (1997). PACT: cloning and characterization of a cellular p53 binding protein that interacts with Rb. *Oncogene*, 14, 145-155.

[12] Li, L., Deng, B., Xing, G., Teng, Y., Tian, C., Cheng, X., Yin, X., Yang, J., Gao, X., Zhu, Y., Sun, Q., Zhang, L., Yang, X., and He, F. (2007). PACT is a negative regulator of p53 and essential for cell growth and embryonic development. *Proc.Natl.Acad.Sci.U.S.A*, 104, 7951-7956.

[13] Sato, S. M. and Sargent, T. D. (1991). Localized and inducible expression of Xenopus-posterior (Xpo), a novel gene active in early frog embryos, encoding a protein with a 'CCHC' finger domain. *Development*, 112, 747-753.

[14] Hoppler, S., Brown, J. D., and Moon, R. T. (1996). Expression of a dominant-negative Wnt blocks induction of MyoD in Xenopus embryos. *Genes Dev.*, 10, 2805-2817.

[15] Wang, H., Gao, M. X., Li, L., Wang, B., Hori, N., and Sato, K. (2007). Isolation, expression, and characterization of the human ZCRB1 gene mapped to 12q12. *Genomics*, 89, 59-69.

[16] Li, H., Liu, Q., Hu, X., Feng, D., Xiang, S., He, Z., Hu, X., Zhou, J., Ding, X., Zhou, C., and Zhang, J. (2009). Human ZCCHC12 activates

AP-1 and CREB signaling as a transcriptional co-activator. *Acta Biochim.Biophys.Sin.(Shanghai)*, 41, 535-544.

[17] Chang, W. L., Lee, D. C., Leu, S., Huang, Y. M., Lu, M. C., and Ouyang, P. (2003). Molecular characterization of a novel nucleolar protein, pNO40. *Biochem.Biophys.Res.Commun.*, 307, 569-577.

[18] Markus, M. A. and Morris, B. J. (2009). RBM4: a multifunctional RNA-binding protein. *Int.J.Biochem.Cell Biol.*, 41, 740-743.

[19] Heo, I., Joo, C., Kim, Y. K., Ha, M., Yoon, M. J., Cho, J., Yeom, K. H., Han, J., and Kim, V. N. (2009). TUT4 in concert with Lin28 suppresses microRNA biogenesis through pre-microRNA uridylation. *Cell*, 138, 696-708.

[20] Rutz, B. and Seraphin, B. (2000). A dual role for BBP/ScSF1 in nuclear pre-mRNA retention and splicing. *EMBO J.*, 19, 1873-1886.

[21] Bai, C. and Tolias, P. P. (1998). Drosophila clipper/CPSF 30K is a post-transcriptionally regulated nuclear protein that binds RNA containing GC clusters. *Nucleic Acids Res.*, 26, 1597-1604.

[22] Hagan, J. P., Piskounova, E., and Gregory, R. I. (2009). Lin28 recruits the TUTase Zcchc11 to inhibit let-7 maturation in mouse embryonic stem cells. *Nat.Struct.Mol.Biol.*, 16, 1021-1025.

[23] Jones, M. R., Quinton, L. J., Blahna, M. T., Neilson, J. R., Fu, S., Ivanov, A. R., Wolf, D. A., and Mizgerd, J. P. (2009). Zcchc11-dependent uridylation of microRNA directs cytokine expression. *Nat.Cell Biol.*, 11, 1157-1163.

[24] Inoue, K., Mizuno, T., Wada, K., and Hagiwara, M. (2000). Novel RING finger proteins, Air1p and Air2p, interact with Hmt1p and inhibit the arginine methylation of Npl3p. *J.Biol.Chem.*, 275, 32793-32799.

[25] Vanacova, S., Wolf, J., Martin, G., Blank, D., Dettwiler, S., Friedlein, A., Langen, H., Keith, G., and Keller, W. (2005). A new yeast poly(A) polymerase complex involved in RNA quality control. *PLoS.Biol.*, 3, e189.

[26] LaCava, J., Houseley, J., Saveanu, C., Petfalski, E., Thompson, E., Jacquier, A., and Tollervey, D. (2005). RNA degradation by the exosome is promoted by a nuclear polyadenylation complex. *Cell*, 121, 713-724.

[27] Wyers, F., Rougemaille, M., Badis, G., Rousselle, J. C., Dufour, M. E., Boulay, J., Regnault, B., Devaux, F., Namane, A., Seraphin, B., Libri, D., and Jacquier, A. (2005). Cryptic pol II transcripts are degraded by a nuclear quality control pathway involving a new poly(A) polymerase. *Cell*, 121, 725-737.

[28] Sanudo, M., Jacko, M., Rammelt, C., Vanacova, S., and Stefl, R. (2011). 1H, 13C, and 15N chemical shift assignments of ZCCHC9. *Biomol.NMR Assign.*, 5, 19-21.

[29] Zhou, A., Zhou, J., Yang, L., Liu, M., Li, H., Xu, S., Han, M., and Zhang, J. (2008). A nuclear localized protein ZCCHC9 is expressed in cerebral cortex and suppresses the MAPK signal pathway. *J.Genet.Genomics*, 35, 467-472.

[30] Kuznicki, K. A., Smith, P. A., Leung-Chiu, W. M., Estevez, A. O., Scott, H. C., and Bennett, K. L. (2000). Combinatorial RNA interference indicates GLH-4 can compensate for GLH-1; these two P granule components are critical for fertility in C. elegans. *Development*, 127, 2907-2916.

[31] Bezares-Calderon, L. A., Becerra, A., Salinas, L. S., Maldonado, E., and Navarro, R. E. (2010). Bioinformatic analysis of P granule-related proteins: insights into germ granule evolution in nematodes. *Dev.Genes Evol.*, 220, 41-52.

[32] Updike, D. and Strome, S. (2010). P granule assembly and function in Caenorhabditis elegans germ cells. *J.Androl*, 31, 53-60.

[33] Spike, C., Meyer, N., Racen, E., Orsborn, A., Kirchner, J., Kuznicki, K., Yee, C., Bennett, K., and Strome, S. (2008). Genetic analysis of the Caenorhabditis elegans GLH family of P-granule proteins. *Genetics*, 178, 1973-1987.

[34] Tzfati, Y., Abeliovich, H., Avrahami, D., and Shlomai, J. (1995). Universal minicircle sequence binding protein, a CCHC-type zinc finger protein that binds the universal minicircle sequence of trypanosomatids. Purification and characterization. *J.Biol.Chem.*, 270, 21339-21345.

[35] Tzfati, Y., Abeliovich, H., Kapeller, I., and Shlomai, J. (1992). A single-stranded DNA-binding protein from Crithidia fasciculata recognizes the nucleotide sequence at the origin of replication of kinetoplast DNA minicircles. *Proc.Natl.Acad.Sci.U.S.A*, 89, 6891-6895.

[36] Milman, N., Motyka, S. A., Englund, P. T., Robinson, D., and Shlomai, J. (2007). Mitochondrial origin-binding protein UMSBP mediates DNA replication and segregation in trypanosomes. *Proc.Natl.Acad.Sci.U.S.A*, 104, 19250-19255.

[37] Sela, D. and Shlomai, J. (2009). Regulation of UMSBP activities through redox-sensitive protein domains. *Nucleic Acids Res.*, 37, 279-288.

[38] Sela, D., Yaffe, N., and Shlomai, J. (2008). Enzymatic mechanism controls redox-mediated protein-DNA interactions at the replication origin of kinetoplast DNA minicircles. *J.Biol.Chem.*, 283, 32034-32044.

[39] Coelho, E. R., Urmenyi, T. P., Franco, d. S., Rondinelli, E., and Silva, R. (2003). Identification of PDZ5, a candidate universal minicircle sequence binding protein of Trypanosoma cruzi. *Int.J.Parasitol.*, 33, 853-858.

[40] Radwanska, M., Couvreur, B., Dumont, N., Pays, A., Vanhamme, L., and Pays, E. (2000). A transcript encoding a proteasome beta-subunit and a zinc finger protein in Trypanosoma brucei brucei. *Gene*, 255, 43-50.

[41] Rajavashisth, T. B., Taylor, A. K., Andalibi, A., Svenson, K. L., and Lusis, A. J. (1989). Identification of a zinc finger protein that binds to the sterol regulatory element. *Science*, 245, 640-643.

[42] Abe, Y., Chen, W., Huang, W., Nishino, M., and Li, Y. P. (2006). CNBP regulates forebrain formation at organogenesis stage in chick embryos. *Dev.Biol*, 295, 116-127.

[43] Chen, W., Liang, Y., Deng, W., Shimizu, K., Ashique, A. M., Li, E., and Li, Y. P. (2003). The zinc-finger protein CNBP is required for forebrain formation in the mouse. *Development*, 130, 1367-1379.

[44] Weiner, A. M., Allende, M. L., Becker, T. S., and Calcaterra, N. B. (2007). CNBP mediates neural crest cell expansion by controlling cell proliferation and cell survival during rostral head development. *J Cell Biochem.*, 102, 1553-1570.

[45] Liquori, C. L., Ricker, K., Moseley, M. L., Jacobsen, J. F., Kress, W., Naylor, S. L., Day, J. W., and Ranum, L. P. (2001). Myotonic dystrophy type 2 caused by a CCTG expansion in intron 1 of ZNF9. *Science*, 293, 864-867.

[46] Niedowicz, D. M., Beckett, T. L., Holler, C. J., Weidner, A. M., and Murphy, M. P. (2010). APP(DeltaNL695) expression in murine tissue downregulates CNBP expression. *Neurosci.Lett.*, 482, 57-61.

[47] Calcaterra, N. B., Armas, P., Weiner, A. M., and Borgognone, M. (2010). CNBP: a multifunctional nucleic acid chaperone involved in cell death and proliferation control. *IUBMB.Life*, 62, 707-714.

[48] Armas, P., Nasif, S., and Calcaterra, N. B. (2008). Cellular nucleic acid binding protein binds G-rich single-stranded nucleic acids and may function as a nucleic acid chaperone. *J.Cell Biochem.*, 103, 1013-1036.

[49] Borgognone, M., Armas, P., and Calcaterra, N. B. (2010). Cellular nucleic-acid-binding protein, a transcriptional enhancer of c-Myc,

promotes the formation of parallel G-quadruplexes. *Biochem.J.*, 428, 491-498.

[50] Armas, P., Aguero, T. H., Borgognone, M., Aybar, M. J., and Calcaterra, N. B. (2008). Dissecting CNBP, a zinc-finger protein required for neural crest development, in its structural and functional domains. *J.Mol.Biol.*, 382, 1043-1056.

[51] Hahn, Y. and Lee, B. (2005). Identification of nine human-specific frameshift mutations by comparative analysis of the human and the chimpanzee genome sequences. *Bioinformatics.*, 21 Suppl 1, i186-i194.

[52] Chureau, C., Prissette, M., Bourdet, A., Barbe, V., Cattolico, L., Jones, L., Eggen, A., Avner, P., and Duret, L. (2002). Comparative sequence analysis of the X-inactivation center region in mouse, human, and bovine. *Genome Res.*, 12, 894-908.

[53] Titulaer, M. K., Mustafa, D. A., Siccama, I., Konijnenburg, M., Burgers, P. C., Andeweg, A. C., Smitt, P. A., Kros, J. M., and Luider, T. M. (2008). A software application for comparing large numbers of high resolution MALDI-FTICR MS spectra demonstrated by searching candidate biomarkers for glioma blood vessel formation. *BMC.Bioinformatics.*, 9, 133.

[54] Scherrer, T., Femmer, C., Schiess, R., Aebersold, R., and Gerber, A. P. (2011). Defining potentially conserved RNA regulons of homologous zinc-finger RNA-binding proteins. *Genome Biol.*, 12, R3.

[55] Sammons, M. A., Samir, P., and Link, A. J. (2011). Saccharomyces cerevisiae Gis2 interacts with the translation machinery and is orthogonal to myotonic dystrophy type 2 protein ZNF9. *Biochem.Biophys.Res.Commun.*, 406, 13-19.

[56] Xu, H. P., Rajavashisth, T., Grewal, N., Jung, V., Riggs, M., Rodgers, L., and Wigler, M. (1992). A gene encoding a protein with seven zinc finger domains acts on the sexual differentiation pathways of Schizosaccharomyces pombe. *Mol.Biol.Cell*, 3, 721-734.

[57] Gobeil, L. A., Plante, P., Rohani, M., Ouellette, M., and Provost, P. (2008). Involvement of Dcr1 in post-transcriptional regulation of gene expression in Schizosaccharomyces pombe. *Front Biosci.*, 13, 2203-2215.

[58] Espinosa, J. M., Portal, D., Lobo, G. S., Pereira, C. A., Alonso, G. D., Gomez, E. B., Lan, G. H., Pomar, R. V., Flawia, M. M., and Torres, H. N. (2003). Trypanosoma cruzi poly-zinc finger protein: a novel DNA/RNA-binding CCHC-zinc finger protein. *Mol.Biochem.Parasitol.*, 131, 35-44.

[59] Park, S. J., Kwak, K. J., Oh, T. R., Kim, Y. O., and Kang, H. (2009). Cold shock domain proteins affect seed germination and growth of Arabidopsis thaliana under abiotic stress conditions. *Plant Cell Physiol*, 50, 869-878.
[60] Park, S. J., Kwak, K. J., Jung, H. J., Lee, H. J., and Kang, H. (2010). The C-terminal zinc finger domain of Arabidopsis cold shock domain proteins is important for RNA chaperone activity during cold adaptation. *Phytochemistry*, 71, 543-547.
[61] Webb, J. R. and McMaster, W. R. (1993). Molecular cloning and expression of a Leishmania major gene encoding a single-stranded DNA-binding protein containing nine "CCHC" zinc finger motifs. *J.Biol.Chem.*, 268, 13994-14002.
[62] McGrath, C. F., Buckman, J. S., Gagliardi, T. D., Bosche, W. J., Coren, L. V., and Gorelick, R. J. (2003). Human cellular nucleic acid-binding protein Zn2+ fingers support replication of human immunodeficiency virus type 1 when they are substituted in the nucleocapsid protein. *J.Virol.*, 77, 8524-8531.
[63] Crooks, G. E., Hon, G., Chandonia, J. M., and Brenner, S. E. (2004). WebLogo: a sequence logo generator. *Genome Res.*, 14, 1188-1190.
[64] Summers, M. F. (1991). Zinc finger motif for single-stranded nucleic acids? Investigations by nuclear magnetic resonance. *J.Cell Biochem.*, 45, 41-48.
[65] South, T. L. and Summers, M. F. (1993). Zinc- and sequence-dependent binding to nucleic acids by the N-terminal zinc finger of the HIV-1 nucleocapsid protein: NMR structure of the complex with the Psi-site analog, dACGCC. *Protein Sci.*, 2, 3-19.
[66] De Guzman, R. N., Wu, Z. R., Stalling, C. C., Pappalardo, L., Borer, P. N., and Summers, M. F. (1998). Structure of the HIV-1 nucleocapsid protein bound to the SL3 psi-RNA recognition element. *Science*, 279, 384-388.
[67] De Guzman, R. N., Turner, R. B., and Summers, M. F. (1998). Protein-RNA recognition. *Biopolymers*, 48, 181-195.
[68] Amarasinghe, G. K., De Guzman, R. N., Turner, R. B., Chancellor, K. J., Wu, Z. R., and Summers, M. F. (2000). NMR structure of the HIV-1 nucleocapsid protein bound to stem-loop SL2 of the psi-RNA packaging signal. Implications for genome recognition. *J.Mol.Biol.*, 301, 491-511.
[69] Guerrerio, A. L. and Berg, J. M. (2010). Design of single-stranded nucleic acid binding peptides based on nucleocapsid CCHC-box zinc-binding domains. *J.Am.Chem.Soc.*, 132, 9638-9643.

[70] Levin, J. G., Guo, J., Rouzina, I., and Musier-Forsyth, K. (2005). Nucleic acid chaperone activity of HIV-1 nucleocapsid protein: critical role in reverse transcription and molecular mechanism. *Prog.Nucleic Acid Res.Mol.Biol.*, 80, 217-286.

[71] Kankia, B. I., Barany, G., and Musier-Forsyth, K. (2005). Unfolding of DNA quadruplexes induced by HIV-1 nucleocapsid protein. *Nucleic Acids Res.*, 33, 4395-4403.

[72] Lyonnais, S., Gorelick, R. J., Mergny, J. L., Le Cam, E., and Mirambeau, G. (2003). G-quartets direct assembly of HIV-1 nucleocapsid protein along single-stranded DNA. *Nucleic Acids Res.*, 31, 5754-5763.

[73] Hamill, S., Wolin, S. L., and Reinisch, K. M. (2010). Structure and function of the polymerase core of TRAMP, a RNA surveillance complex. *Proc.Natl.Acad.Sci.U.S.A*, 107, 15045-15050.

In In: Zinc Fingers
Editors: R. Ciofani and L. Makrlik
ISBN 978-1-62100-230-7


*Chapter 3*

# Designing HIV-RREIIB RNA Binding Zinc Fingers

***Markus W. Germann,[1,2,*] Subrata H. Mishra,[3] Yoshiko Santoso[1] and Alexander M. Spring[1]***
[1]Departments of Chemistry
[2]Biology and the Neuroscience Institute, Georgia State University
[3]Biophysics and Biophysical Chemistry
School of Medicine, Johns Hopkins University

## Abstract

RNA is central to many cellular and viral functions. Agents that interact with specific RNAs offer means of intervention and serve as tools for functional studies. We have demonstrated the successful targeting of an HIV RNA substructure (RREIIB) by zinc finger proteins. These proteins were initially evolved by phage display, subsequently pruned to a single zinc finger and further enhanced by a structure and thermodynamic guided design. The structure of the zinc finger and the changes on the RNA target induced upon binding were investigated using an array of techniques including gel shift experiments, NMR spectroscopy and isothermal titration calorimetry. In addition to the α-helical portion, other secondary structures of the zinc finger also

[*]E-mail: mwg@gsu.edu, Phone: 404-413-5561, Fax: 404-413-5505

contribute to the specific binding. Insights from the thermodynamic analysis provide a detailed understanding of the molecular processes of the interaction and highlighted the crucial role of water in the binding process. Taking the effect of favorable interactions and water into account, zinc fingers with higher RREIIB-Tr binding affinity were generated. Our findings suggest strategies for enhancing the binding affinities of other biomolecular systems.

## Introduction

Zinc fingers are small compact protein structural motifs folded through coordination of at least one zinc ion by various combinations of cysteine and histidine side chains. [1-4] The zinc finger motif accounts for 1-3 % of human genes and is a widely used functional protein domain in eukaryotes. [5, 6] Zinc fingers (Znf) interact with DNA, RNA, proteins, lipids and small molecules to affect cellular processes such as transcription, chromatin remodeling, DNA repair, and apoptosis. RNA specific zinc fingers also exhibit a similar diversity (i.e. 3' mRNA processing, poly-A binding proteins, RNA nucleocytoplasmic transport, ribosomal structural proteins, and post-transcriptional regulation of the p53 gene). [7-18] Infectious organisms also utilize zinc finger proteins to regulate their aberrant RNA in a cellular environment. The HIV-1 nucleocapsid protein (NC), for example, facilitates the strand transfer process between the viral RNA and newly reverse-transcribed DNA, and is also required for proper viral genome packaging. [19, 20]

Since their discovery in the mid 80's, several zinc finger classes with different biological functions have been reported.[18, 21] They can be classified based on the zinc coordinating amino acids as CCHH, CCHC, CCCH and CCCC zinc fingers.[16, 21] By taking the arrangements of zinc coordinating amino acids and secondary structures into account, 8 fold groups were described. [18] The most common of these, the CCHH zinc finger class of approximately 30 amino acids, is characterized by an N-terminal antiparallel β-sheet and a C-terminal α-helix stabilized by a zinc ion forming a ββα fold.[22] Additional stabilization comes from three conserved large hydrophobic residues (-12, -3 and +4, based on the start of the α-helix) that form the hydrophobic core of the zinc finger.[4] The zinc ion acts as a Lewis acid and accepts 4 electron pairs.[23] These are provided by two cysteines and two histidines that coordinate the free zinc ion tetrahedrally and displace 6 water molecules. [24, 25] Under physiological conditions, the histidine (imidazole nitrogen) coordinating the zinc ion is uncharged while the cysteines are deprotonated. [26] (Figure 4C) Zinc coordination with a thiol group is possible and is more likely for higher numbers of cysteine ligands.[27, 28]

The design of ZnF's that target DNA reliably has developed into a mature field. This success has sparked an interest to also develop target specific RNA binding zinc fingers to alter and intervene with RNA dependent cellular functions. In this article we review the targeting of HIV RREIIB RNA by single and dual ZnFs that were originally generated by phage display. Subsequently, specific mutations were introduced to assess their impact on structure and thermodynamics of RNA recognition and to improve the affinity to the target RNA.[29-32]

# Zinc Fingers Targeting Nucleic Acids

*DNA vs RNA targets:* Although formed from similar building blocks, DNA and RNA present quite different binding interfaces for zinc finger proteins. DNA is predominately double stranded and tends be regular in structure with a wide and shallow major groove. [Figure 1a] When bound to DNA, CCHH zinc fingers are situated in the major groove making specific contacts between α-helix side chains and typically 3 sequential DNA bases allowing for sequence specific recognition. Further stabilization derives from electrostatic phosphate contacts and non-specific hydrophobic interactions. [4, 33, 34] Zinc fingers are quite often found in proteins as tandem repeats that are connected by short linkers, most often of the Krüppel-type TGEKP. The repeats allow for higher sequence specificity while the linkers aid in overall stabilization of the DNA-ZnF complex via proper spacing of the fingers.[4, 35, 36]

In contrast, regular double helical RNA presents a narrow but deep major groove that cannot accommodate an α-helix.[37] [Figure 1B] However, RNA exhibits a wide variety of secondary and tertiary structures with bulges, loops, extrahelical bases, and junctions that can be interrogated by an α-helix. [38, 39] CCHH zinc finger binding of double stranded helical RNA typically occurs at or near bulges and loops that feature widened grooves and unique structural features.[37] Contacts are generally made with the phosphodiester backbone although some sequence specificity can occur through hydrophobic stacking interactions with extrahelical or bulged bases.[40] The short linkers between tandem ZnF repeats serve a similar purpose as with the DNA binding motifs: proper orientation of the zinc finger.[36] Overall the ZnF-RNA contacts are specific for the spatial presentation of the RNA structure. Since the RNA structure is sequence dependent, this corresponds to an indirect sequence readout.

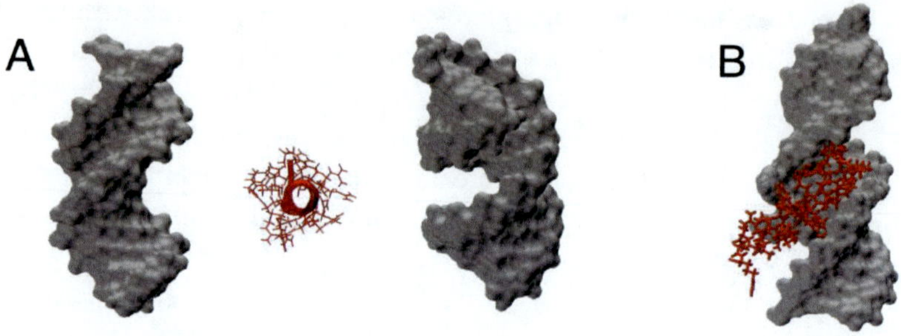


Figure 1. A) Surface representations of canonical B-type DNA (left) and A-type RNA (right) with a stick representation of the α-helical rev peptide between the two (pdb ID: 1etf)[40]. Major grooves of DNA and RNA are facing towards the rev peptide. B) Solution NMR structure of RREIIB-Tr RNA (grey) with the rev peptide (red) bound along the major groove face of the bulge region. Note how the α-helix is accommodated within the widened bulge (pdb 1etf)[40].

Interestingly, unlike CCHH, the CCCH motif is known to bind to ssRNA in a sequence specific manner and is linked to activities associated with post transcriptional regulation of gene expression and nucleocytoplasmic RNA shuttling.[13, 15, 41, 42]

*The TFIIIA Protein:* The original and perhaps most extensively studied zinc finger protein is the transcription factor IIIA (TFIIIA) of *Xenopus laevis.* Intriguingly, TFIIIA serves as a DNA binding transcription factor for the 5S gene and an RNA binding cofactor of the 5S ribosomal RNA and binds both with similarly high $K_D$'s (500 – 700 pM).[1, 42, 43] The TFIIIA protein consists of 9 tandem CCHH zinc finger motifs. All are involved in DNA binding; fingers 1 – 3 are crucial for binding while some (4 and 6) just serve as spacer elements. Similar to DNA binding, all 9 zinc fingers are utilized in RNA binding with fingers 4 – 6 being the minimum needed.[44]. Of particular interest for RNA specific zinc finger design are fingers 4 and 6 that make base and backbone contacts with a purine rich bulge and a highly ordered loop region of 5s rRNA respectively but only function as spacer elements when bound to DNA.[38, 41, 42]

# HIV-1, Rev, and the Rev Response Element (RRE)

The Human Immunodeficiency Virus-1 (HIV-1) rev protein (a non-zinc finger protein) binds to a highly conserved specific RNA structure called the rev responsive element (RRE) in HIV introns.[45, 46] Rev binding to RRE is absolutely required for the production of virions. In early stages of HIV mRNA expression, the transcribed viral RNA undergoes multiple splice events before nucleocytoplasmic transport. This RNA encodes for various HIV-1 regulatory proteins, including rev, which predominantly localizes in the nucleus.

In later stages of mRNA expression, as rev accumulates, it binds as a monomer to the IIB loop of RRE (RREIIB). Subsequent multimerization binding to the RRE inhibits splicing and signals for cytoplasm transport. The unspliced and singly spliced viral RNA encodes the viral structural proteins as well as provides unspliced genomic RNA for virion packaging. This event ushers in the late phase of the cellular HIV infection: virion assembly and budding. [47-51] The importance of the rev-RRE interaction for HIV-1 replication makes it an obvious target for therapeutic intervention by molecular means.

# Generation of RRE Binding Peptides

*Rev peptides:* The RNA binding region of rev has been isolated to an arginine rich sequence of the protein with a high specificity for the purine rich bulge in RREIIB ($K_D$ of 1 to 3 nM).[49] Peptides based on this sequence express the same specificity and have been shown to adopt an α-helical conformation when bound. Solution NMR structures of the rev peptide have been solved with various versions of the RREIIB RNA construct including a truncated version, RREIIB-Tr.[48, 52] (Figure 1B and 2).

Complex formation between rev and RREIIB-Tr causes the formation of two GC and two purine-purine (GG, GA) base pairs in the bulge, which are critical to rev binding [47]. Subsequent studies of the rev peptide have identified key side-chain interactions required for binding. In particular, Asn40 forms hydrogen bonds to the major groove face of the GA base pair in the bulge region.[40, 53] Other studies showed that a higher α-helical predisposition of the peptide improved binding.[54]

Expanding on this, the Frankel lab has shown that inclusion of a modified version of the RRE binding rev peptide sequence into a zinc finger motif maintains a ββα fold and exhibits RREIIB specificity.[37] An alternative approach is to employ phage display techniques to optimize the proteins sequence of zinc fingers to effect specific binding to RREIIB.[29]

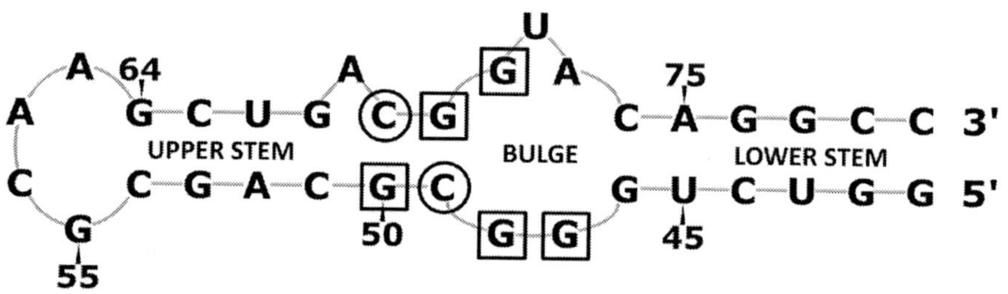


Figure 2. RREIIB-Tr sequence with named regions. The numbering corresponds to RREIIB RNA. Circles represent C to U mutations and squares G to 2-aminopurine mutations for the thermodynamic profiling of the RNA.

*Phage display:* The stable and versatile ZnF framework has been subjected to biocombinatorial, computational and protein engineering strategies.[2, 29] Phage display allows the screening of the effect of multiple amino acid changes on a protein simultaneously. The protein library is displayed on the surface of a bacteriophage and selection is based on binding of the displayed protein to a desired target. Proteins, and hence phages, that do not bind are eliminated by washing while those that bind are eluted, amplified in bacteria, isolated and subjected to additional selection cycles as desired. Members of the library that have the required properties are identified from sequencing the phage DNA.[29, 55]

An HIV RREIIB RNA binding tandem zinc finger construct was prepared based on the 4$^{th}$ zinc finger of TFIIIA. The sequence library encompassed amino acids between positions -2 and +10 of the α-helix keeping the zinc coordinating histidines and amino acids that contribute to the hydrophobic core unchanged; the beta sheet was not altered. These constructs were displayed on the surface of bacteriophage fd as a coat protein III fusion, selected for binding to RREIIB RNA, enriched and subsequently sequenced. One abundant construct named RR1 specifically bound the RRE-IIB target with a $K_D$ of 7.9 nM.[29] Examination of the amino acid sequence of RR1 showed that the first zinc finger RR1-zf1 was acidic while the second zinc finger was basic. When these zinc fingers were prepared separately the second zinc finger RR1-zf2 bound RREIIB RNA with a $K_D$ of 26 nM demonstrating that specific binding is due to the second basic zinc finger component of RR1.[32]

The potential to utilize a single zinc finger to specifically bind the RNA target is attractive because of their remarkable stability, resistance to proteolytic degradation and small size. These qualities facilitate diffusion and are expected to assist nuclear localization.

*Zinc finger like peptides:* In addition to optimizing and pruning RREIIB binding zinc finger constructs to the smallest functional size as was done for ZNF29, a descendent of RRI-zf2, we also explored the possibility of generating peptide structures that maintain RNA binding without requiring zinc while possibly maintaining a zinc finger like structure (Table 1).

Peptides were designed that do not support zinc coordination, instead stabilization was introduced by using hydrophobic or ionic contacts. This approach is based on a previous successful report where the hydrophobic core of a similar construct was reconfigured.[56, 57] Several candidate peptides were designed and evaluated using the program FoldX, and subsequently a subset was probed for structural integrity by running 50ns accelerated molecular dynamics simulations in water.[58, 59] (Table 1) Sequences that maintained an overall zinc finger structure at the end of the simulation were selected and synthesized. The peptides were evaluated by circular dichroism and NMR spectroscopy and found to be partially unstructured under normal solution conditions (unpublished results). Nevertheless, the peptides were still able to bind the RNA target as demonstrated by a fluorescence displacement assay (Figure 3, Table 1).

Two of the peptides were subsequently characterized by gel shift assays. In this series the FDFK peptide exhibit highest affinity as judged from the fluorescence displacement and the gel shift assays, demonstrating relatively high affinity to HIV RREIIB-Tr RNA in the absence of the zinc (Table 1). However, the requirement and challenge to engineer a stable scaffold limits the freedom to change residues. In addition, the high stability of the zinc fingers is a distinct advantage for evaluating details of the RNA protein binding.

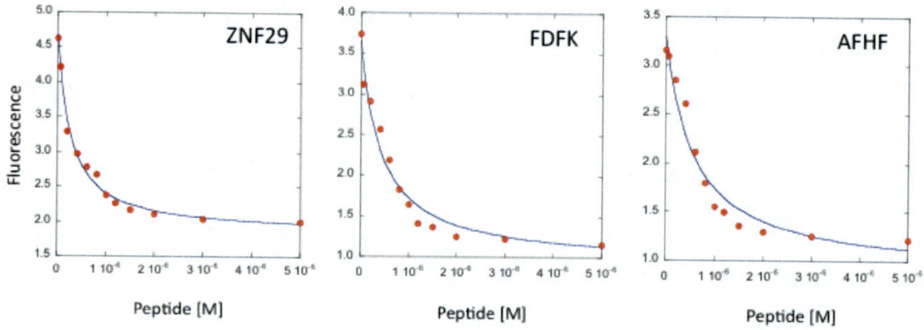


Figure 3. Ethidium bromide fluorescence displacement assay of zinc finger ZNF29 and selected peptides. Fluorescence was recorded with excitation at 505 nm and emission at 602 nm for 50nM RREIIB-TR RNA, 50nM ethidium bromide, 10 mM sodium phosphate, 50 mM NaCl at pH 7.0 in a 0.5 cm cuvette at room temperature.

**Table 1. Accelerated MD was run for 50ns at 300K using Amber 9 and the parm99 force field. Protein charges were neutralized with chloride ions and solvated with TIP3 water (15Å). A boost was applied to the dihedral torsions (the total dihedral angle obtained +100, e.g = 277.75 kcal/mol + 100 kcal/mol).[59] RMSD were compared to the starting structures and given for the entire molecule or just the β-sheet. $K_D$ represents dissociation constants obtained from a gel shift assay (50 nM RNA, 10 mM sodium phosphate, 50 mM NaCl). $K_I$ values represent the amount of peptide needed to effect 50% reduction of the ethidium bromide fluorescence (50nM RREIIB-Tr RNA and 50nM ethidium bromide, 10 mM sodium phosphate, 50 mM NaCl at pH 7.0 and ambient temperature)**

| Name | Sequence | RMSD | | $K_I$[μM] | $K_D$ [μM] |
|---|---|---|---|---|---|
| | | all | β-sheet | | |
| ZNF29 | MVYV**C**HFEN**C**GRSFNDRRKLNR**H**KKI**H**TG | 2.7 | 1.4 | 0.2 | - |
| ZNF29 no $Zn^{2+}$ | MVYV**C**HFEN**C**GRSFNDRRKLNR**H**KKI**H**TG | 9.2 | 8.4 | 1.2 | - |
| FDFK | MVYV**F**HFEN**D**GRSFNDRRKLNR**F**KKI**K**TR | 1.5 | 1.4 | 0.5 | 0.30 |
| AFHF | MVYV**A**HFEN**F**GRSFNDRRKLNR**H**KKI**F**TR | 2.3 | 2.4 | 0.6 | 0.48 |
| LLFF | MVYV**L**HFEN**L**GRSFNDRRKLNR**F**KKI**F**TR | 3.0 | 1.3 | 0.7 | - |
| FFFF | MVYV**F**HFEN**F**GRSFNDRRKLNR**F**KKI**F**TR | 2.2 | 2.0 | 1.3 | - |
| KLDL | MVYV**K**HFEN**L**GRSFNDRRKLNR**D**KKI**L**TR | 2.9 | 2.9 | - | - |
| LLLL | MVYV**L**HFEN**L**GRSFNDRRKLNR**L**KKI**L**TR | 3.5 | 2.6 | - | - |
| DDKK | MVYV**D**HFEN**D**GRSFNDRRKLNR**K**KKI**K**TR | 3.9 | 3.1 | - | - |
| DDRR | MVYV**D**HFEN**D**GRSFNDRRKLNR**R**KKI**R**TR | 4.0 | 3.9 | - | - |
| VVLL | MVYV**V**HFEN**V**GRSFNDRRKLNR**L**KKI**L**TR | 3.9 | 3.6 | - | - |

# Structure and Thermodynamics of RRE Binding Zinc Fingers

## 1. Solution Structure of ZNF29 and ZNFG29R

The observation that a single zinc finger is sufficient for high affinity targeting of HIV RREIIB RNA greatly simplifies the analysis of important amino acid residues and detailed binding studies.[32] In addition to ZNF29 that was derived from the RR1-zf2 protein, several additional mutants were created to examine interactions with RREIIB (discussed in the next section). To employ a structure based design approach to construct RNA binding proteins, we focused on the structure of monomeric zinc finger proteins ZNF29 and ZNFG29R.[31] A mutant of ZNF29, ZNFG29R containing a terminal arginine instead of a glycine was designed to enhance RNA binding.

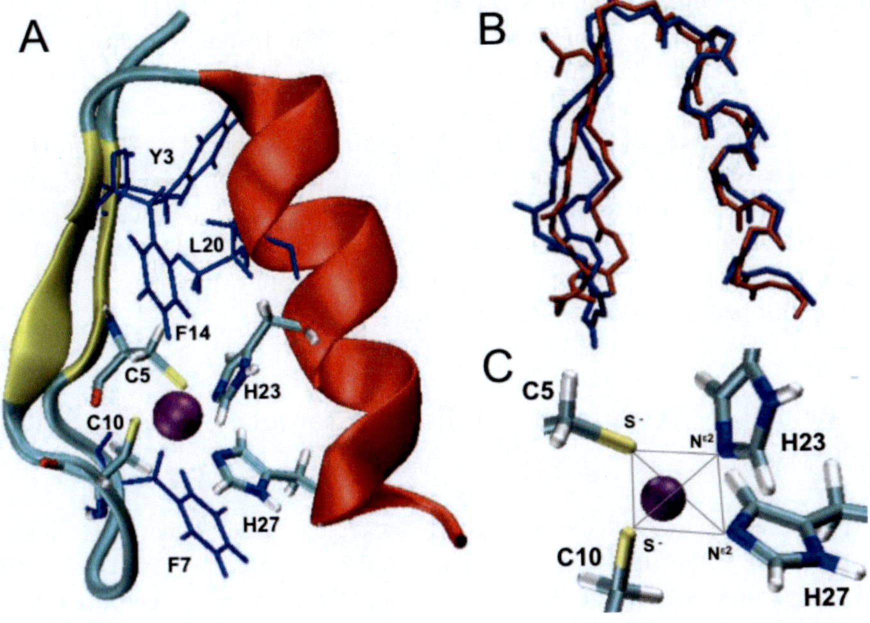


Figure 4. (A) NMR solution structure of ZNF29 (pdb id: 2AB3)[31] displaying the zinc coordinating residues (Cys5, Cys10, His23 and His27) and other residues that constitute the hydrophobic core (Tyr3, Phe7, Phe14, Leu20) of the zinc finger. (B) An overlay of the protein backbones of ZNF29 (blue) (residues 2-28) and the 4$^{th}$ zinc finger of TFIIIA (red) (pdb id: 1TF6, residues 104-130).[63] The RMSD is 1.68 Å. (C) Tetrahedral coordination of the $Zn^{2+}$ ion (purple) in ZNF29 by the thiolates of Cys5 and Cys10 and the $N^{\varepsilon 2}$ nitrogens of His23 and His27.

The ββα fold of the CCHH class of zinc finger proteins results in diagnostic upfield chemical shifts for the conserved Phe ζH in the hydrophobic core and His δ2H involved in $Zn^{2+}$ coordination.[60] These resonances greatly facilitate monitoring of folded and unfolded states of the expressed protein, which is important for both structural and binding studies. In the absence of zinc, both ZNF29 and ZNFG29R unfold and the chemical shifts of these resonances return to a random coil value. Once folded however they exhibited stability over a wide temperature and pH range.[31] Both ZNF29 and ZNFG29R have the characteristic ββα architecture of CCHH zinc finger proteins and are structurally closely related to the TFIIIA template that was used for the phage display (Figure 4B).

The hydrophobic core consists of Tyr3, Phe7, Phe14, Leu20 and the zinc coordinating residues of Cys5, Cys10, His23 and His27 (Figure 4A). The amino acids Tyr3 – Cys5 and Arg12 – Phe14 constitute the two β-strands of the antiparallel β-sheet. Between the β-sheet and a 3 turn α-helix (Arg17-Thr28) is the tip of the finger containing an Asn15 and Asp16. The RMSD between the backbones of ZNF29 and ZNFG29R (residues 2-28) is 1.03Å indicating that the substitution of glycine for arginine at the C terminal position does not perturb the structure.

The structures account for the upfield shifted peaks (Phe14 ζH and His27 δ2H) referred to previously due to ring current effects. The zinc coordinating histidines are in the neutral tautomeric $NH^{\delta_1}$ form, which implicates the $N^{\varepsilon_2}$ nitrogen in zinc coordination; this is consistent with the chemical shifts of these nitrogens.[31] The presence of metal coordinating cysteines as thiolates was confirmed from $Co^{2+}$ substitution studies.[61, 62]

## 2. Binding of Zinc Fingers to the RNA Target Studied by NMR and ITC

The energetics of the molecular processes that facilitate RNA–ZnF recognition were studied by isothermal titration calorimetry (ITC) and NMR. ITC experiments delineate the effects of salt, pH, temperature and the role of solvent in binding. Titration of RREIIB-Tr with ZNF29 monitored by the NMR chemical shift changes in the imino proton region for the RNA, revealed a different mode of binding than the rev peptide yet retains a 1:1 binding stoichiometry.[31, 48] Specifically, the imino proton spectra of the ZnF bound RREIIB-Tr RNA has only 3 new imino proton peaks indicating the formation of fewer new base pairs as compared to the rev peptide binding to the same

RNA. Similar to the rev peptide, base pairs flanking the RNA bulge (U66 – A52, U45 – A75) display prominent chemical shift changes while base pairs in the stem regions are unaffected. Taken together this localizes the binding to the RNA bulge, which was also confirmed for all the ZnF mutants making the comparison of their thermodynamic binding parameters possible. Single mutations to ZNF29 were introduced on the β-sheet (H6A, H6K, R12A), tip (N15A, D16A) and the α–helix (N21A, N21Q, G29R). Individual contributions of amino acids and RNA base pairs to binding were extracted from selected ZnF and RNA variants using ITC.[30] (Figure 5B,C) As anticipated from the NMR experiments, the binding isotherms can be fit to a single binding site model with a 1:1 stoichiometry; relevant thermodynamic parameters ($\Delta H$, $\Delta S$, $K_D$) are listed in Table 2 and Figure 5 B,C. The free energy of binding at 25°C for all ZNF29 mutants – RREIIB-Tr interactions has favorable enthalpy and entropy contributions. Changes in binding affinity discussed below for the ZnF mutants are based on the control, RREIIB-Tr/ZNF29 interaction.

*ZnF β sheet mutations:* Histidine 6 is important for RNA binding because a change to alanine (H6A) lowers the affinity 1.4 fold while the lysine mutant (H6K) has a 1.6 fold increased binding. From a combination of NMR and ITC experiments it was determined that the pKa of H6 increases from 6.7 to 8.3 in the bound form. Therefore the negatively charged local binding environment of H6 stabilizes a protonated histidine, which potentially primes it for interaction with the phosphodiester backbone. The R12A mutation on the second β-strand of ZNF29 causes a 1.8 fold reduction in the binding affinity. A comparison of the affinity of ZNF29 and ZNF29R12A for RNA under increasing salt concentrations provides evidence that R12 makes an electrostatic contact with the RNA phosphate backbone. The involvement of amino acids located in the β-sheet is obviously different from the binding mode of the rev peptide that only contains α-helical residues. Both H6 and R12 flank the β-sheet and may serve to anchor the ZnF to the RNA in coordination with other interactions in the widened RNA major groove. These interactions provide opportunities for additional optimization strategies.

*Mutations at the tip of the zinc finger:* This region of the zinc finger is critical for RNA binding. Both ZNFN15A and ZNFD16A result in the lowest binding affinities (2.2 and 3 fold lower) amongst all ZnF mutants studied. In addition, the new imino proton peaks that appear on complex formation of these two mutants with RREIIB-Tr, display distinct chemical shift differences as compared to the control (ZNF29). These results indicate that N15 and D16 are important for structural features that support the formation of new base pairs in the complex.

**Table 2. Binding affinities are expressed as dissociation constants ($K_D$) where $K_D = 1/K_A$, the association constant obtained by curve-fitting the ITC binding isotherm to a 1:1 site binding model**

| Protein | RNA | $K_D$ (nM) |
|---|---|---|
| ZNF29 | RREIIB-Tr | 149 |
| H6A | RREIIB-Tr | 207 |
| H6K | RREIIB-Tr | 95 |
| R12A | RREIIB-Tr | 266 |
| N15A | RREIIB-Tr | 322 |
| D16A | RREIIB-Tr | 450 |
| N21A | RREIIB-Tr | 232 |
| N21Q | RREIIB-Tr | 80 |
| G29R | RREIIB-Tr | 57 |
| N21QG29R | RREIIB-Tr | 35 |
| | | |
| G29R | G50_2AP | 210 |
| G29R | C69U | 254 |
| G29R | G70_2AP | 671 |
| G29R | C49_U | 510 |
| G29R | G48_71_2Ap | 324 |
| G29R | G47_2AP | 303 |

*ZnF α helix mutations:* Asn21 of ZNF29 could potentially serve a similar role as Asn40 of the rev peptide that contacts a newly formed G47-A73 base pair in the complex. The N21A and N21Q mutants had opposite effects on the RNA binding; the N21A mutation reduces binding affinity 1.6 fold while there is a 1.9 fold increase for N21Q. The side chain of Asn21 does not serve the same role as Asn40 in the rev peptide because, the imino proton spectra of ZNF29N21A-RREIIB-Tr is similar to that of the control. Finally, the G29R mutation at the C-terminus of the helix increases RNA binding by 2.6 fold. Based on the salt dependence, Arg29 was determined to make an electrostatic contact with the RNA phosphodiester backbone.

In conclusion, the ITC studies indicate that amino acid positions at all secondary structure elements of the zinc fingers are directly involved in the binding interaction.

*RNA mutations:* RNA variants were used to probe the effects of base pair disruption on binding. Guanosines in the middle stem (G50 and G70) and the bulge (G48-G71 and G47) were replaced with 2-aminopurine (ap) and cytosines in the middle stem (C49, C69) with uracil. (Figure 2) The modified RNAs bind ZNFG29R with 1:1 stoichiometry and have favorable enthalpy and entropy contributions at 25 °C except for the RNA mutant with a 2-aminopurine substitution at position G70 (G70_ap), which has favorable enthalpy but unfavorable entropy contribution.

While all the RNA mutants display lower binding affinities than the wild type RNA (Table 2), a trend was observed depending on the position of the mutation. The G70 – C49 base pair is not formed in the free RNA as determined from NMR; however, any changes disrupting the potential G70–C49 base pair in the middle stem upon binding causes a drastic reduction in ZnF binding affinity (~12 fold lower for G70_ap and ~9 fold lower for C49U). The 2-aminopurine substitutions at the bulge (G48G71_ap, G47_ap) reduce ZnF binding affinity ~ 5 - 6 fold while alteration of the G50 – C69 base pair have the least deleterious effect on ZnF binding. The trend in the binding enthalpies for RNA variants illustrates that specific base pair formation in the bulge and in the immediate vicinity of the bulge (G70 –C49) is crucial for the configuration of the ZnF binding site.

## Influence of Water on the Binding Thermodynamics

A crucial component in biomolecular interactions that is frequently ignored is the participation of water. Attention was drawn to the involvement of water by the observation that a) mutations with higher binding affinity did not also exhibit a more exothermic binding enthalpy; the $\Delta H_{obs}$ of ZNFG29R binding is lower than ZNF29 (Figure 5 B) despite the additional favorable electrostatic interaction of the terminal arginine with the phosphate backbone, and b) the temperature dependence of the binding enthalpy (i.e. the heat capacity of binding, $\Delta C_p$) is substantially different for the 2 ZnF's (-403 ± 40 and -274 ± 34 cal mol$^{-1}$ K$^{-1}$ for ZNF29 and ZNFG29R respectively).

Generally, a non-zero binding heat capacity results from major structural rearrangements of the biomolecules upon binding.[64] This is, however, not the case for either the RNA or ZnF since free and complexed RNA show only minor differences in the CD spectra and the chemical shifts of the residues in the ZnF hydrophobic core are similar for the free and bound ZnF (ZNF29 and ZNFG29R).[31] Moreover, there is a considerable difference in $\Delta C_p$ of binding for the 2 ZnF's even though both proteins effect the same changes on the

RNA. The observed $\Delta C_p$ values cannot be rationalized from conventional models since these correlate large negative values to extensive dehydration of mainly non-polar surfaces at the binding interface.[65, 66] For ZNF29 and ZNFG29R-RNA binding, more polar than non-polar surface is buried. Similar observations have also been reported for other biomolecular interactions.[67, 68]

Large negative values of $\Delta C_p$ for biomolecular interactions that cannot be explained by surface dehydration have been rationalized in an alternative model proposed by Alan Cooper.[69] In this model, trends in $\Delta H$ and $\Delta C_p$ are attributed to the extent of hydrogen bond changes resulting from differences in ligand induced water displacement at the binding site. (Figure 5A) The model makes the following predictions:

(1) Binding enthalpy of a ligand that displaces more water molecules than another ligand is less exothermic.
(2) Water involvement can result in negative $\Delta C_p$. Specifically, a ligand displacing less water on binding is predicted to have a more negative $\Delta C_p$. Each additional water molecule displaced increases $\Delta C_p$ by +18 cal $mol^{-1}K^{-1}$.

ZNFG29R displaces more water than ZNF29 on binding the solvated RNA bulge; consequently, the observed binding enthalpies are in agreement with this model. Similarly, as predicted, $\Delta C_p$ of ZNF29 binding is more negative than that of ZNFG29R.

Based on the difference in heat capacity, ZNFG29R displaces ~7 more water molecules than ZNF29. This is in good agreement with the number estimated (~7 to 8) from the difference in their solvent excluded volumes (110 $Å^3$). Similar trends became apparent for other ZNF29 mutants where despite favorable interactions a less exothermic enthalpy was recorded in agreement with the Cooper model (Figure 5B).

Essentially, a mutation to smaller non polar amino acid has 2 opposing effects on the enthalpy: an enthalpic loss due to the potential absence of a favorable interaction, compensated by an enthalpic gain due to lower displacement of water. The overall affinity is further modulated by a lower entropic gain when less bound water is released. These effects are also observed for the RNA variants where fewer displaced water molecules again result in a gain in enthalpy and a reduction of the binding entropy (Figure 5C).

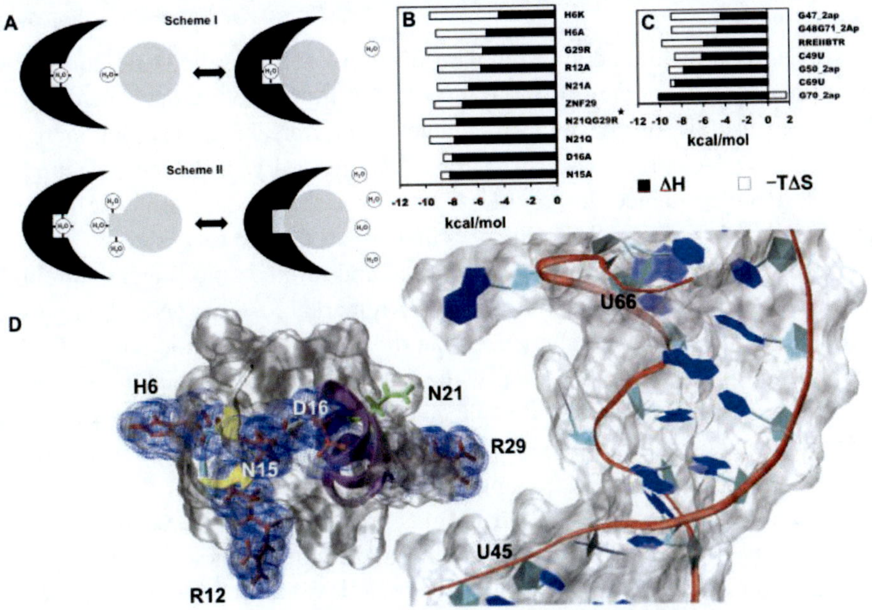


Figure 5. (A) Schematic representation of the Cooper model. The ligand (grey) in scheme 2 displaces more water than ligand in scheme 1 on binding the same macromolecular cavity (black), adapted from [69]. Schemes 1 and 2 represent binding of ZNF29 and ZNFG29R respectively. (B) Enthalpic and entropic contributions to the binding free energy at 298 K for ZNF29 mutants binding RREIIB-Tr and (C) for RREIIB-Tr mutants binding ZNFG29R. The binding enthalpies are displayed in the order of increasing exothermicity. Entropic contributions are expressed as –TΔS. All ZnF mutants are single point mutants except for N21QG29R (marked with ★). (D) A representation of ZNFG29R as it encounters RREIIB-Tr (pdb id: 1ETF) major groove at the bulge. ZnF residues that are expected to be positioned at the protein-RNA-solvent interface in the complex are rendered with blue dots (His6, Arg12, Asn15, Asp16, Arg29). The interaction at the Asn21 position will be secluded from the solvent in the complex. The phosphates of the RNA residues U66 and U45 are at the entrance to the major groove; the chemical shift of the imino protons of these residues change on ZnF binding.

Analysis of the mutants indicates that the local binding interface at N21 does not involve water. Hence the difference in binding enthalpies of the N21 mutants do not contain solvent effects and are expected to reflect only the potential presence (N21Q) or absence (N21A) of a favorable interaction (Figure 5B). This information has been successfully used to design a double mutant ZNFN21QG29R, where both the enthalpic and entropic contributions to free energy are favorably increased as compared to ZNF29 and exhibits the highest binding affinity of all the mutant proteins.

# Conclusion

RNA protein interactions regulate essential cellular and viral functions, including mRNA splicing, RNA transport and translation. Consequently it is of interest to develop means to modulate or interfere with these interactions. Targeting DNA with zinc finger proteins generally involves selecting a specific DNA sequence. In contrast RNA targeting is more cumbersome, less developed and dependent on structures that are accessible to a protein.

Using phage display we have generated and refined zinc fingers with high specificity for HIV RREIIB RNA. It is remarkable that ZNF29 targets a completely different RNA sequence than the RNA target of the phage display template despite having essentially the same backbone architecture (Figure 4B). This demonstrates that the phage display generated ZNF29 does not contain any unusual properties and that the essentially rigid zinc finger scaffold can be successfully manipulated to recognize a variety of RNA's. The high stability of the ZnF scaffold permits the variation of a wide variety of residues, provided that the zinc chelating and hydrophobic core residues are not compromised. The prospect of using "zinc less" zinc finger like structures is appealing since the oxidation and modification of the zinc coordinating residues or replacement of zinc may inactivate or unfold the ZnF motif. However, peptides investigated in this study to date, are clearly structurally much less robust although they still recognize the RNA target.

Although only the α-helix and the tip were subjected to randomization in the phage display, ZNF29 uses the full array of secondary structures for interaction with the RREIIB target including the β-sheet residues. This provides additional opportunities to harness β-sheet and loop residues in addition to the obvious α-helical residues for combinatorial screening. Given the high affinity of the ZnF's for the target, it may become difficult to select proteins with higher affinity using phage display. It may then necessitate the use of zinc finger competitor to winnow the phage population to the highest affinity members.

The thermodynamic analysis in conjunction with the structural studies highlights the importance of water. The enthalpy and entropy contributions of displaced water upon binding need to be taken into account when designing mutants with seemingly favorable interactions. Mutating sites of interactions where water is not present simplifies the analysis and does not obscure the effect of favorable interactions. These strategies were successfully applied in the design of a zinc finger double mutant with superior binding characteristics.

Such an approach may be instrumental in improving affinity of drugs guided by thermodynamic studies. In addition to targeting and blocking access to specific RNA structures, such zinc fingers have the potential to be developed into molecular tools by adding task specific protein domains similar to their DNA specific counterparts.

## Acknowledgments

This work was supported by grants from the NIH, the Georgia Cancer Coalition, the Georgia State University Brains and Behavior and Molecular Basis of Disease fellowships. We are indebted to David Wilson for many stimulating discussions.

## References

[1] Miller, J.; A.D. McLachlan; A. Klug. *The EMBO Journal*. 1985. 4(6), 1609- 1614.

[2] Klug, A. *Annu Rev Biochem*. 2010. 79, 213-31.

[3] Brown, D.D. *Cell*. 1984. 37(2), 359-65.

[4] Iuchi, S. C2H2 Zinc Fingers as DNA Binding Domains. In *Zinc Finger Proteins: From Atomic Contact to Cellular Function*; S. Iuchi and N. Kudell; Ed.; Landes Bioscience / Eurekah.com and Kluwer Academic / Plenum Publishers: New York, NY, 2005, pp. 7 - 13.

[5] Henikoff, S.; E.A. Greene; S. Pietrokovski; P. Bork; T.K. Attwood; L. Hood. *Science*. 1997. 278(5338), 609-14.

[6] Chen, Y.; G. Varani. *FEBS J*. 2005. 272(9), 2088-97.

[7] Mandel, C.R.; Y. Bai; L. Tong. *Cell Mol. Life Sci*. 2008. 65(7-8), 1099-122.

[8] Prouteau, M.; M.-C. Daugeron; B. Séraphin. *EMBO J*. 2008. 27(22), 2966-76.

[9] Hurt, J.A.; R.A. Obar; B. Zhai; N.G. Farny; S.P. Gygi; P.A. Silver. *The Journal of Cell Biology*. 2009. 185(2), 265-77.

[10] Kelly, S.M.; S.W. Leung; L.H. Apponi; A.M. Bramley; E.J. Tran; J.A. Chekanova; S.R. Wente; A.H. Corbett. *J. Biol. Chem*. 2010. 285(34), 26022-32.

[11] Varmeh-Ziaie, S.; I. Okan; Y. Wang; K.P. Magnusson; P. Warthoe; M. Strauss; K.G. Wiman. *Oncogene*. 1997. 15(22), 2699-704.

[12] Chen, T.; A.M. Brownawell; I.G. Macara. *Mol. Cell Biol.* 2004. 24(15), 6608-19.

[13] Lai, W.S.; P.J. Blackshear. *J. Biol. Chem.* 2001. 276(25), 23144-54.

[14] Wang, Y.; L. Opperman; M. Wickens; T.M.T. Hall. *Proceedings of the National Academy of Sciences*. 2009. 106(48), 20186.

[15] Ling, A.S.; J.R. Trotter; E.F. Hendriks. *Journal of Biological Chemistry*. 2011.

[16] Laity, J.H.; B.M. Lee; P.E. Wright. *Curr. Opin. Struct. Biol.* 2001. 11(1), 39-46.

[17] Nakamura, H.; H. Kawagishi; A. Watanabe; K. Sugimoto; M. Maruyama; M. Sugimoto. *Molecular and Cellular Biology*. 2011.

[18] Krishna, S.S.; I. Majumdar; N.V. Grishin. *Nucleic Acids Res.* 2003. 31(2), 532-50.

[19] Bazzi, A.; L. Zargarian; F. Chaminade; C. Boudier; H. De Rocquigny; B. Rene; Y. Mely; P. Fosse; O. Mauffret. *Nucleic Acids Res*. 2011.

[20] D'Souza, V.; M.F. Summers. *Nat. Rev. Microbiol.* 2005. 3(8), 643-55.

[21] Darby, M.K. RNA Binding by Single Zinc Fingers. In *Zinc Finger Proteins: From Atomic Contact to Cellular Function*; S. Iuchi and N. Kudell; Ed.; Landes Bioscience / Eurekah.com and Kluwer Academic / Plenum Publishers: New York, NY, 2005, pp. 66 - 75.

[22] Berg, J.M. *Annu. Rev. Biophys. Biophys. Chem.* 1990. 19, 405-21.

[23] Williams, R.J.P. *Polyhedron*. 1987. 6(1), 61 - 69.

[24] Larabee, J.L.; J.R. Hocker; J.S. Hanas. *Chem. Res. Toxicol.* 2005. 18(12), 1943-54.

[25] Dudev, T.; C. Lim. *J. Am. Chem. Soc.* 2000. 122(45), 11146 - 11153.

[26] Blasie, C.A.; J.M. Berg. *Biochemistry*. 2002. 41(50), 15068-73.

[27] Fabris, D.; Y. Hathout; C. Fenselau. *Inorg. Chem.* 1999. 38(6), 1322-1325.

[28] Aras, S.; G. Singh; K. Johnston; T. Foster; A. Aiyar. *PLoS Pathog.* 2009. 5(6), e1000469.

[29] Friesen, W.J.; M.K. Darby. *Nat. Struct. Biol.* 1998. 5(7), 543-6.

[30] Mishra, S.H.; A.M. Spring; M.W. Germann. *J. Mol. Biol.* 2009. 393(2), 369-82.

[31] Mishra, S.H.; C.M. Shelley; D.J. Barrow, Jr.; M.K. Darby; M.W. Germann. *Biopolymers*. 2006. 83(4), 352-64.

[32] Friesen, W.J.; M.K. Darby. *J. Biol. Chem.* 2001. 276(3), 1968-73.

[33] Pavletich, N.P.; C.O. Pabo. *Science*. 1991. 252(5007), 809-17.

[34] Klug, A. The Discovery of Zinc Fingers and Their Practical Application in Gene Regulation: A Personal Account. In *Zinc Finger Proteins: From Atomic Contact to Cellular Function*; S. Iuchi and N. Kudell; Ed.; Landes Bioscience / Eurekah.com and Kluwer Academic / Plenum Publishers: New York, NY, 2005, pp. 1 - 6.
[35] Brown, R.S.; J. Flint. TFIIIA: A Sophisticated Zinc Finger Protein. In *Zinc Finger Proteins: From Atomic Contact to Cellular Function*; S. Iuchi and N. Kudell; Ed.; Landes Bioscience / Eurekah.com and Kluwer Academic / Plenum Publishers: New York, NY, 2005, pp. 14 - 19.
[36] Ryan, R.F.; M.K. Darby. *Nucleic Acids Res*. 1998. 26(3), 703-9.
[37] McColl, D.J.; C.D. Honchell; A.D. Frankel. *Proc. Natl. Acad. Sci. U S A*. 1999. 96(17), 9521-6.
[38] Brown, R.S. *Curr. Opin. Struct. Biol*. 2005. 15(1), 94-8.
[39] Stefl, R.; L. Skrisovska; F.H. Allain. *EMBO Rep*. 2005. 6(1), 33-8.
[40] Battiste, J.L.; H. Mao; N.S. Rao; R. Tan; D.R. Muhandiram; L.E. Kay; A.D. Frankel; J.R. Williamson. *Science*. 1996. 273(5281), 1547-51.
[41] Hall, T.M.T. *Curr. Opin. Struct. Biol*. 2005. 15(3), 367-73.
[42] Romaniuk, P.J. TFIIIA and p43. *In Zinc Finger Proteins: From Atomic Contact to Cellular Function*; S. Iuchi and N. Kudell; Ed.; Landes Bioscience / Eurekah.com and Kluwer Academic / Plenum Publishers: New York, NY, 2005, pp. 56 - 65.
[43] Klug, A.; D. Rhodes. *Cold Spring Harb. Symp. Quant. Biol*. 1987. 52, 473-82.
[44] Pieler, T.; O. Theunissen. *Trends Biochem. Sci*. 1993. 18(6), 226-30.
[45] Hope, T.; R.J. Pomerantz. *Curr. Top Microbiol. Immunol*. 1995. 193, 91-105.
[46] Pollard, V.W.; M.H. Malim. *Annu. Rev. Microbiol*. 1998. 52, 491-532.
[47] Battiste, J.L.; R. Tan; A.D. Frankel; J.R. Williamson. *J. Biomol. NMR*. 1995. 6(4), 375-89.
[48] Battiste, J.L.; R. Tan; A.D. Frankel; J.R. Williamson. *Biochemistry*. 1994. 33(10), 2741-7.
[49] Brice, P.C.; A.C. Kelley; P.J. Butler. *Nucleic Acids Res*. 1999. 27(10), 2080-5.
[50] Edgcomb, S.P.; A. Aschrafi; E. Kompfner; J.R. Williamson; L. Gerace; M. Hennig. *Protein Sci*. 2008. 17(3), 420-30.
[51] Daelemans, D.; S.V. Costes; E.H. Cho; R.A. Erwin-Cohen; S. Lockett; G.N. Pavlakis. *J. Biol. Chem*. 2004. 279(48), 50167-75.
[52] Patel, D.J. *Curr Opin Struct Biol*. 1999. 9(1), 74-87.

[53] Ye, X.; A. Gorin; A.D. Ellington; D.J. Patel. *Nat. Struct. Biol.* 1996. 3(12), 1026-33.
[54] Tan, R.; L. Chen; J.A. Buettner; D. Hudson; A.D. Frankel. *Cell.* 1993. 73(5), 1031-40.
[55] Smith, G.P. *Science.* 1985. 228(4705), 1315-7.
[56] Dahiyat, B.I.; S.L. Mayo. *Science.* 1997. 278(5335), 82-7.
[57] Sarisky, C.A.; S.L. Mayo. *J. Mol. Biol.* 2001. 307(5), 1411-8.
[58] Schymkowitz, J.; J. Borg; F. Stricher; R. Nys; F. Rousseau; L. Serrano. *Nucleic Acids Res.* 2005. 33(Web Server issue), W382-8.
[59] Hamelberg, D.; J. Mongan; J.A. McCammon. *J. Chem. Phys.* 2004. 120(24), 11919-29.
[60] Lee, M.S.; R.J. Mortishire-Smith; P.E. Wright. *FEBS Lett.* 1992. 309(1), 29-32.
[61] Frankel, A.D.; J.M. Berg; C.O. Pabo. *Proc. Natl. Acad. Sci. U S A.* 1987. 84(14), 4841-5.
[62] Mishra, S.H., Structure and Energetics of RNA - Protein Interactions for HIV RREIIB Targeting Zinc Finger Proteins, Department of Chemistry, 2008, Georgia State University: Atlanta.
[63] Nolte, R.T.; R.M. Conlin; S.C. Harrison; R.S. Brown. *Proc. Natl. Acad. Sci. U S A.* 1998. 95(6), 2938-43.
[64] Privalov, P.L.; I. Jelesarov; C.M. Read; A.I. Dragan; C. Crane-Robinson. *J. Mol. Biol.* 1999. 294(4), 997-1013.
[65] Spolar, R.S.; M.T. Record, Jr. *Science.* 1994. 263(5148), 777-84.
[66] Spolar, R.S.; J.R. Livingstone; M.T. Record, Jr. *Biochemistry.* 1992. 31(16), 3947-55.
[67] Bergqvist, S.; M.A. Williams; R. O'Brien; J.E. Ladbury. *J. Mol. Biol.* 2004. 336(4), 829-42.
[68] Holdgate, G.A.; A. Tunnicliffe; W.H. Ward; S.A. Weston; G. Rosenbrock; P.T. Barth; I.W. Taylor; R.A. Pauptit; D. Timms. *Biochemistry.* 1997. 36(32), 9663-73.
[69] Cooper, A. *Biophys Chem.* 2005. 115(2-3), 89-97.

In In: Zinc Fingers
Editors: R. Ciofani and L. Makrlik
ISBN 978-1-62100-230-7


*Chapter 4*

# ZNF224: Structural and Functional Insights into a KRAB Zinc Finger Protein

***Angelo Lupo[1], Paola Izzo[2] and Paola Costanzo[3]***

[1]Department of Biological and Environmental Sciences, Faculty of Sciences, University of Sannio Benevento Italy

[2]Department of Biochemistry and Medical Biotechnology, Faculty of Medicine, University of Naples "Federico II" Italy

[3]Department of Biochemistry and Medical Biotechnology, Faculty of Biotechnology, University of Naples "Federico II" Italy

## Introduction

During evolution, the origin of new characteristics is accompanied by the appearance of novel genes. These genes, namely regulatory genes, can modulate effector genes located downstream in the functional pathways. Therefore, quantitative and qualitative modifications of these regulatory genes result in impressive changes in several biological processes. New biological functions also occur through the production of new protein variants. The modular structure of these proteins has allowed them to develop the capability to interact with DNA and RNA and with other protein partners, thus turning on or off regulatory pathways. A combination of ancient and modern modules can originate a complex protein repertoire bringing about the loss or gain of new specific functions in the various animal species. Proteins able to play these regulatory roles, in most cases, belong to the transcription factor family.

Transcription factors can activate or repress gene transcription through binding to the promoter region of the target genes. Regulation of the transcription rate often occurs through protein-protein interactions that modify the properties of the transcription factors, causing gene activation or repression. Finally, transcription factors also regulate DNA accessibility by interacting with various coactivator or corepressor proteins that are capable of modifying the chromatin structure in the promoter region of target genes.

KRAB-containing zinc finger proteins (KRAB-ZFPs), a remarkable subfamily of the most widespread C2H2 zinc finger protein (C2H2-ZNF) family, belong to the third type of transcription regulators described above. In fact, they contain a powerful transcriptional repression domain at the NH2-terminus, called KRAB (*Krüppel*-Associated Box), which recruits various activities modifying the chromatin structure around the promoter region of target genes. The transcriptional repression mediated by the KRAB-ZFPs plays a crucial role in gene expression, cell proliferation, differentiation, apoptosis and cancer.

ZNF224, a member of the KRAB-ZFPs, is a paradigmatic example of a protein made up of different modules and, consequently, displays multiple functions. This chapter will focus on the structure, expression and biological functions of ZNF224.

## Evolution and Structure of the ZNF224 Gene

The *Krüppel*-type, or C2H2-containing zinc finger (ZNF) proteins (C2H2-ZNFs) represent one of the most ancient and numerous protein families. They are encoded by genes widely scattered in all eukaryotic genomes. The C2H2 zinc finger genes constitute about 2% of the entire human genome and most of these genes encode for transcription factors that bind to the DNA motifs through the zinc finger domains, thus modulating gene transcription positively or negatively [1].

Within the C2H2-ZNF family, which includes about 800 members, a homogeneous group of proteins have been identified which, besides the zinc finger motifs, show additional domains that can specifically assign new functions to them [2]. In fact, at the NH2 terminus of these proteins the KRAB (*Krüppel*-Associated Box), the SCAN (SRE-ZBP, CTfin51, AW-1 and Number18 cDNA) and the BTB (Broad complex Tramtrack Bric-a-brac) domains have been found [3-5]. While the SCAN domain is involved in mediating protein-protein interaction and the BTB motif acts as a dimerization domain, the KRAB domain plays a crucial role in repressing gene transcription. The model of repression foresees that the KRAB-ZFPs bind to their corresponding DNA sequence through the zinc finger domains, while the KRAB domain binds to the KAP-1 corepressor. KAP-1 functions as a scaffold protein, namely recruiting on the promoters of target genes histone deacetylases, methylases and ATP-dependent activities that modify the chromatin, silencing gene expression through the formation of a facultative heterochromatin environment [6-12].

SCAN and KRAB domains are vertebrate-specific, whereas the BTB motif is also present in insects. In particular, the KRAB repression module is specific of the genomes of vertebrate tetrapods. It is absent in the zinc finger proteins deriving from fish, *Drosophila*, plants, fungi and yeast, whereas it has been found in the frog, chicken, rat, mouse and human genomes [2, 13]. This expression profile suggests that the KRAB domain is a rather recent product of evolution and that the expansion of the KRAB-ZFPs in vertebrate tetrapods is due to new functions related to differentiation and speciation.

Most of the C2H2-ZNF genes are distributed across all chromosomes in the human genome, whereas the KRAB-ZFPs and SCAN-ZFPs are clustered in selected positions on the chromosomes. To date, at least 65 clusters for the KRAB-ZFPs and SCAN-ZFPs have been identified in humans; 384 genes encoding for KRAB-ZFPs are arranged in these clusters. One hundred of these genes are conserved in mammals, and their orthologs in various species have been found. In humans, the majority of genes encoding for KRAB-ZFPs are included in 25 major clusters, the largest of which is positioned on chromosome 19 [14, 15].

The ZNF224 gene is also mapped at the human gene locus 19q13.2 and has no orthologs in mice, suggesting that events of duplication and loss of common ancestral genes occurred independently after the divergence of the lineages in humans and mice [14].

A comparison between ZNF224 genomic sequences and cDNA shows that the coding region is divided into six exons (Figure 1A). All exon/intron boundaries follow the classic GT/AG rule. Fusion of exons I, II and III is responsible for the formation of a 303 bp-long 5'UTR, whereas the complete coding region and a short 3'UTR are generated by the fusion of exons IV, V and VI [16]. In detail, exons IV and V, respectively, encode for the A and b boxes of the KRAB domain, whereas the numerous zinc finger modules are generated by the single large exon VI, as observed in other KRAB-ZFPs.

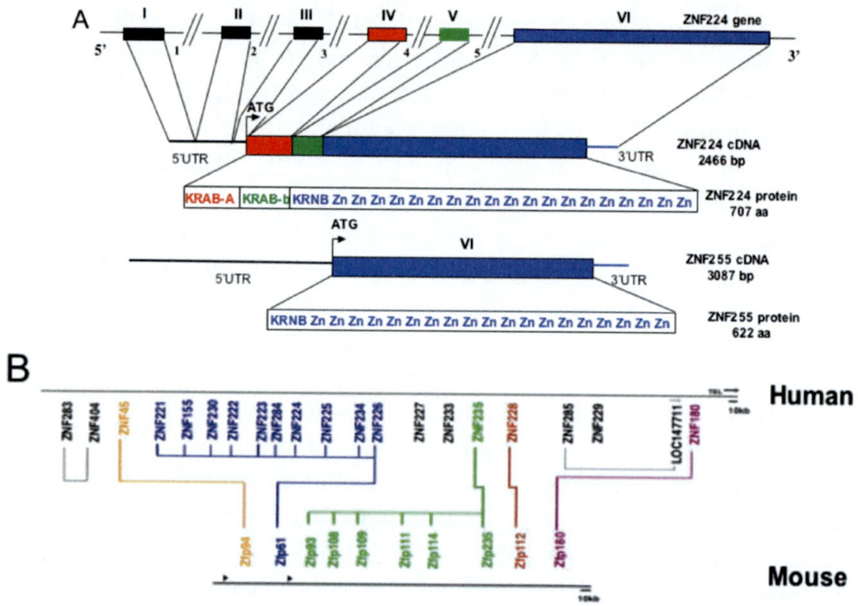


Figure 1. (A) Genomic structure of ZNF224.The exons are indicated by Roman numerals, the introns by Arabic numbers. ZNF224 and ZNF255 cDNAs are shown. ATG indicates the translation start sites. The two proteins originating from the different mRNA isotypes are shown. (B) Organization of the orthologs and paralogs of the KRAB-ZNFs in the human and mouse maps. The 700-bp long fragment includes 21 genes belonging to the human ZNF gene family. The 300 bp-long region comprises 10 members of the mouse gene family. Lines of different colors connect genes present in both gene families. The lines indicate also the groups of genes where a single gene in one species belongs to a clade of different genes in the other group. The human genes reported in black are not connected because the relationships with the counterpart in the mouse are not clear. The arrows indicate the positions of the genes and the two black triangles represent two degenerate KRAB and zinc finger domains. Reproduced with permission from Shannon et al. Differential expansion of zinc finger transcription factor loci in homologous human and mouse gene clusters. Genome Research, (2003) 13, 1097-1110. CSHL Press.

The ZNF224 gene encodes for a 707 amino acid-long protein that contains 19 tandemly-repeated C2H2-type zinc finger motifs separated by a highly conserved spacer sequence of seven amino acids (TGEKPYX), referred to as the H/C link [16]. Analysis of the NH2-terminus sequence of ZNF224 shows a highly conserved 45 amino acid-long KRAB A box and a degenerate 30 amino acid-long KRAB b box (Figure 1A) [16].

On the basis of the primary structure of the KRAB domain, the mammalian KRAB-ZFPs can be divided into three groups: those that include an A box alone (the KRAB A family), those that contain both A and B boxes (the KRAB A+B family) and those that show a canonical A box and degenerate b box, also called C box [2]. An additional regulatory domain, named *Krüppel*-related novel box (KRNB), showing a transcriptional repression activity, is also shown in Figure 1A [17, 18]. The fact that the KRAB-ZFPs are organized in structurally independent modules is probably one of the reasons behind the expansion of this protein family in humans.

Moreover, the notion that in the human genome, 423 independent KRAB-ZFP-coding loci may generate at least 742 structurally distinct proteins by alternative splicing mechanisms or different promoter usage led us to speculate that alternative transcripts could also be generated by the single ZNF224 gene [15].

Identification and characterization of a second transcript, namely ZNF255, definitively strengthened this hypothesis [19]. This additional mRNA, which is 3087 bp-long, bears a longer 5'UTR (1080 bp) than ZNF224 and the same 3'UTR. Through the translation of this transcript, a new 622 amino acid-long protein product is generated. This protein, spanning the region between amino acids 86 and 707 of ZNF224, lacks the KRAB domain, but contains the 19 zinc finger modules [19]. Therefore, two protein species, ZNF224 and ZNF255, showing a different NH2-terminus (with the presence or absence of the KRAB domain), originate from a single gene (Figure 1A). These data are in accordance with the results observed in other KRAB-containing ZFPs [20, 21].

## Differential Expression of the ZNF224 Gene

In recent evolution, a rapid expansion of the mammalian KRAB-ZNF gene family has been generated by duplication of common ancestral genes, leading to more than 400 genes in the human genome. On the other hand, large deletions have occurred in the mouse genome and caused the loss of entire gene clusters in this species. Both events, which happened after the divergence between primate and rodent lineages, turned out to be part of a powerful mechanism to originate new genes during evolution and, consequently, generate new putative and species-specific functions [13, 14]. Human chromosome 19 contains a large number of the KRAB-ZNF genes clustered at 11 major sites. The ZNF224 gene is included in a cluster on the human chromosome 19 (19q13.2), which contains another 20 different genes. The analogous cluster present on the mouse chromosome 7, on the contrary, contains only 10 genes [13]. Figure 1B schematically shows the organization of the orthologs and paralogs in the genomic regions of the human chromosome 19 and mouse chromosome 7, respectively.

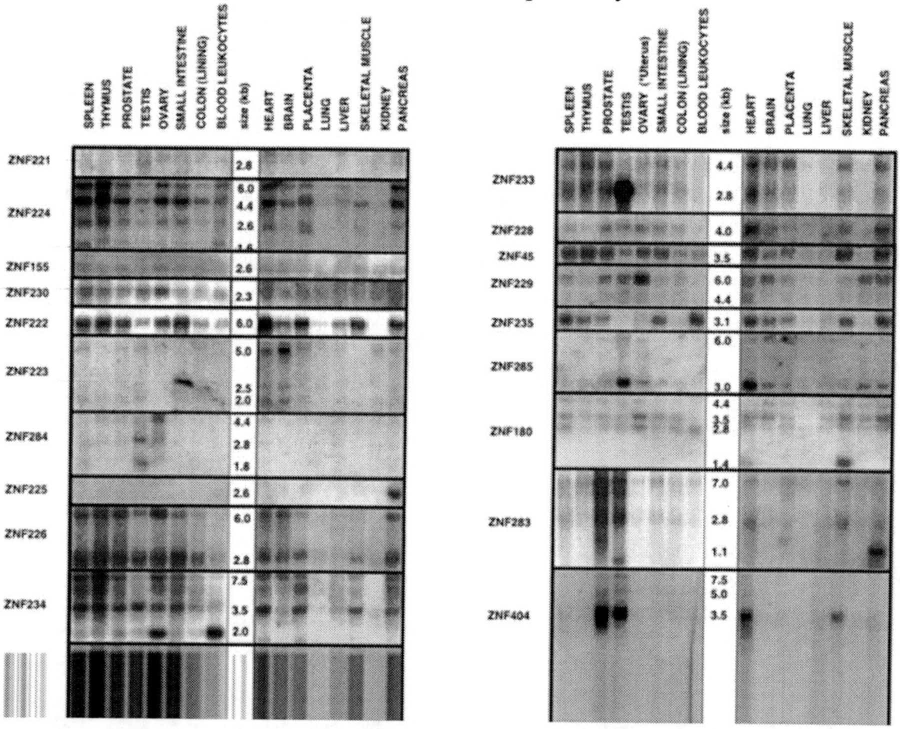


Figure 2. Expression of the KRAB-ZNFs located in position HSA19q13.2 in different human tissues. Northern blotting analysis was carried out using specific probe to hybridize to the poly (A)+ mRNA extracted from the indicated human tissues. The names of the genes reported in the figure are grouped in accordance with the evolutionary relationships between the loci. The middle panels show the molecular weights of the transcripts. Reproduced with permission from Shannon et al. Differential expansion of zinc finger transcription factor loci in homologous human and mouse gene clusters. Genome Research, (2003) 13, 1097-1110. CSHL Press.

It is known that genes encoding olfactory receptor (OLFR) and Kruppel-type-containing ZNF proteins are particularly prone to familial clustering, but while the genes encoding for olfactory receptors are, coherently, co-regulated and co-expressed, thus indicating that they share common regulatory regions, on the contrary, the KRAB-ZFP genes, although similarly organized in clusters, are not co-regulated (22, 23). Most of the KRAB-ZFP genes, which are clustered with ZNF224 at 19q13.2, are expressed in a ubiquitous manner in adult tissues (Figure 2). However, within the cluster located on 19q13.2, which includes the ZNF224 gene, there are also other genes, closely positioned to each other (ZNF225, ZNF284 and ZNF223) that are expressed only in few tissues. In addition, some of these genes (ZNF222, ZNF230, ZNF155, ZNF221 and ZNF225) express only a single mRNA species, whereas other genes (ZNF223, ZNF284, ZNF224, ZNF234 AND ZNF226) belonging to the same group generate mRNA variants showing different lengths (Figure 2). ZNF224 mRNA is ubiquitously produced in various tissues (brain, heart, skeletal muscle, thymus, spleen, liver, lung, leukocytes and lymphocytes). As mentioned above, at least two distinct mRNAs, differing in sequence length, have derived from the single ZNF224 gene. To study the differential expression of the two mRNA isoforms, various cDNAs extracted from fetal and adult human tissues were analyzed by semiquantitative RT-PCR, using sequence-specific oligonucleotides designed to discriminate between ZNF224- and ZNF255-specific transcripts.

The results shown in Figures 3A and B demonstrate that, whereas ZNF224 isoform is expressed in all tissues tested, ZNF255, on the contrary, is more widely expressed in the fetal than in the adult tissues analyzed. These data represent the first experimental evidence of the existence of two isoform mRNAs differing in the 5' UTR and suggest that from these mRNAs two distinct proteins showing a different NH2-terminus originate [19].

In proliferating Caco-2, IMR90 and HeLa human cells, ZNF224 is highly expressed and its ability to repress gene transcription is strongly activated. However, in the same cells (Caco-2, IMR90, HeLa) differentiated, or arrested by serum deprivation, the expression of ZNF224 is significantly reduced, resulting in the removal of repression of target gene transcription (data not shown). The alternating occupancy of the aldolase A promoter by the ZNF224 repression complex in proliferating and growth-arrested cells suggests that this complex plays a significant role during the cell cycle modulation of the aldolase A gene expression [24]. Further findings on new target genes whose transcription is modulated by ZNF224 could be crucial to explain the emerging role of this KRAB-ZFP in the regulation of cell cycle-dependent gene transcription.

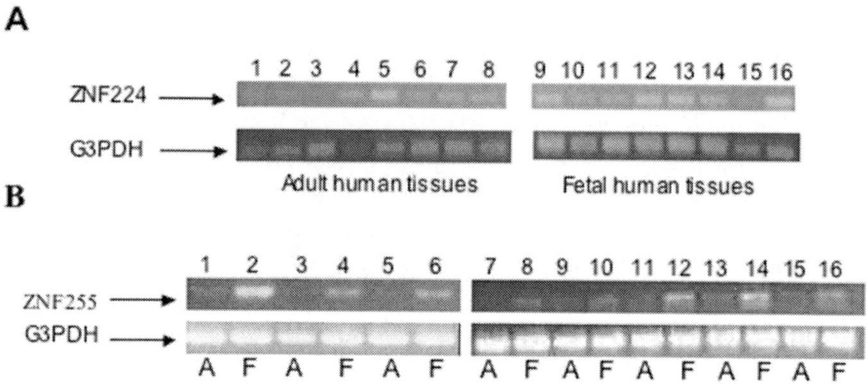


Figure 3. PCR analysis of ZNF224 and ZNF255 transcripts. (A) Equal amounts of cDNAs extracted from adult (lanes 1-8) and fetal (lanes 9-16) human brain, heart, skeletal muscle, thymus, spleen, kidney, liver and lung were used in PCR assays in presence of the specific oligonucleotides for ZNF224. The upper arrow indicates the 365 bp-long ZNF224 amplification product, while the 440 bp-long G3PDH amplification product is shown as a comparative PCR reaction control. (B) Equal amounts of cDNAs extracted from fetal human brain, lung, kidney, heart, skeletal muscle, thymus, spleen and liver (lanes 2, 4, 6, 8, 10, 12, 14, 16) and from adult human brain, lung, kidney, heart, skeletal muscle, thymus, spleen and liver (lanes 1, 3, 5, 7, 9, 11, 13, 15) were used in PCR assays in presence of the specific oligonucleotides for ZNF255. The upper arrow indicates the 310 bp-long ZNF255 amplification product, while the 440 bp-long G3PDH amplification product is shown as a comparative PCR reaction control. A and F indicate adult and fetal tissue, respectively. Reproduced with permission from Medugno et al. Differential expression and cellular localization of ZNF224 and ZNF255, two isoforms of the Krüppel-like zinc-finger protein family. (2007) Gene 403, 125-131. Elsevier.

## Differential Cellular Localization and Functions of ZNF224 and ZNF255

In the previous section, it was stated that two different protein species, ZNF224 and ZNF255, originate from the single ZNF224 gene. The main difference between these two proteins is the presence or absence of the KRAB domain. Biochemical and functional studies demonstrated that ZNF224 acts as a transcriptional repressor of the human aldolase A gene and that ZNF224-mediated repression requires the specific interaction of the KRAB-A domain with the KAP1 corepressor molecule and the involvement of histone deacetylase activity, as is well known in other cases of KRAB-mediated gene silencing [16, 25].

An additional component of the ZNF224 repressor complex recently identified is arginine methyltransferase PRMT5 [26]. It was demonstrated that the methylation of arginine 3 of histone H4 by PRMT5 occurs on the nucleosomes surrounding the aldolase A promoter region. This chromatin modification is associated with the repression of aldolase A gene expression, suggesting a novel mechanism of chromatin structure regulation that is required for the repression of gene transcription by KRAB-ZFPs. The association of ZNF224/PRMT5 repression complex with the aldolase A promoter and the methylation status of H4R3 both vary during the cell cycle, confirming that ZNF224 plays a significant role in the cell-cycle modulation of the aldolase A gene expression. Moreover, our finding of global changes in PRMT5 expression and symmetric H4R3 dimethylation during the cell cycle lead us to speculate that PRMT5-mediated methylation of histones is required to elicit a repression of the genes whose function is correlated to the alternating phases of the cell cycle [26]. Our data are in agreement with the recently demonstrated role played by PRMT5 in the transcriptional repression of cell-cycle regulator and tumor suppressor genes. [27, 28].

While ZNF224 is known to be a transcriptional repressor, the biological function of ZNF255 isoform is still largely unknown. In agreement with the notion that the KRAB domain is a strong repression domain, the ZNF255 isoform, which lacks the KRAB, exhibits a reduced repression activity compared with the ZNF224 isoform [19]. Moreover, the cellular distribution of the ZNF224 and of the ZNF255 isoforms appear to be quite different, as seen by immunofluorescent staining of the Flag-tagged fusion proteins expressed in various human cell types. The ZNF224 protein is present only in the nucleus and shows a diffuse nuclear distribution, typical of a transcription factor. On the contrary, ZNF255 is localized in both the nucleus and cytoplasm (Figure 4A) and is not uniformly distributed in the nucleus, but displays a speckled nuclear distribution, a pattern commonly associated with nuclear splicing factors. Western blotting analysis performed on nuclear and cytoplasmic extracts confirms this different subcellular distribution of ZNF224 and ZNF255 (Figure 4B). An intriguing finding revealed that ZNF255 is present in the nucleolus, as emerged by immunofluorescent staining (Figure 4C), where it is shown to co-localize with fibrillarin, an abundant nucleolar protein. This observation, combined with the speckled distribution of ZNF255, could suggest a role for ZNF255 in splicing; indeed, other components with established roles in mRNA splicing have also been shown to be associated with the nucleolus before integration into the spliceosomes [29]. However, the nucleolar localization of ZNF255 could have functional implications for ribosomal biosynthesis. Treatment of HeLa cells with a low dosage of actinomycin D (0.05 μg/ml), which usually inhibits RNA polymerase I [28], caused translocation of ZNF255 to the nucleolar periphery (Figure 4C), suggesting that ZNF255 could be involved in the transcription of rRNA genes and in the processing of rRNA into preribosomal particles [30]. So far, only a few Kruppel-type zinc-finger proteins have been reported to associate with the nucleoli, but the biological significance of this localization remains elusive [31-33].

ZNF224 and ZNF255 seem to display typical features of the KRAB-ZF proteins that originate from a single gene by alternative splicing or different promoter usage, giving rise to multiple isoforms with distinct subcellular localizations and strong functional differences [19]. The suggestion of a divergent function of ZNF224 and ZNF255 is strengthened by the identification of their specific pattern of interaction with the Wilms' tumor protein (WT1), a zinc finger transcription factor involved in the development of the urogenital system and of numerous other tissues/organs [34].

WT1 can act as an oncogene in leukemia and various types of solid tumors, primarily through the inhibition of apoptosis [35, 36]. The WT1 protein has two major isoforms, WT1(+KTS) and WT1(–KTS), because of the insertion or exclusion of three amino acids (KTS) between the third and fourth zinc fingers; WT1(–KTS) appears to act mainly as a transcriptional factor, while WT1(+KTS) is involved in RNA processing [37]. Interestingly, ZNF224 co-localizes and shows a specific nuclear interaction with the WT1 transcriptional isoform (-KTS), while ZNF255 co-localizes and interacts with both WT1 isoforms, with a preference for the (+KTS) isoform (Figure 5 A-H). These interactions are designed to control different cellular pathways.

In fact, by cotransfection and chromatin immunoprecipitation experiments using a known WT1 target gene, the vitamin D receptor (VDR), it was demonstrated that ZNF224 acts as a transcriptional co-activator of WT1 on the VDR promoter, providing the first example of a KRAB-ZFP that does not act simply as a transcriptional repressor [38]. Conversely, ZNF255 colocalizes with WT1 in translating ribosomes and is present in ribonuclear protein (RNP) complexes, thus suggesting that ZNF255 may participate with WT1(+KTS) in RNA maturation and post-transcriptional control [38]. The isoform-specific interactions of ZNF224/255 with an ancient ZF transcription factor such as WT1 suggest that they have evolved in a specific way to interact with specific isoforms of a protein conserved throughout vertebrate radiation.

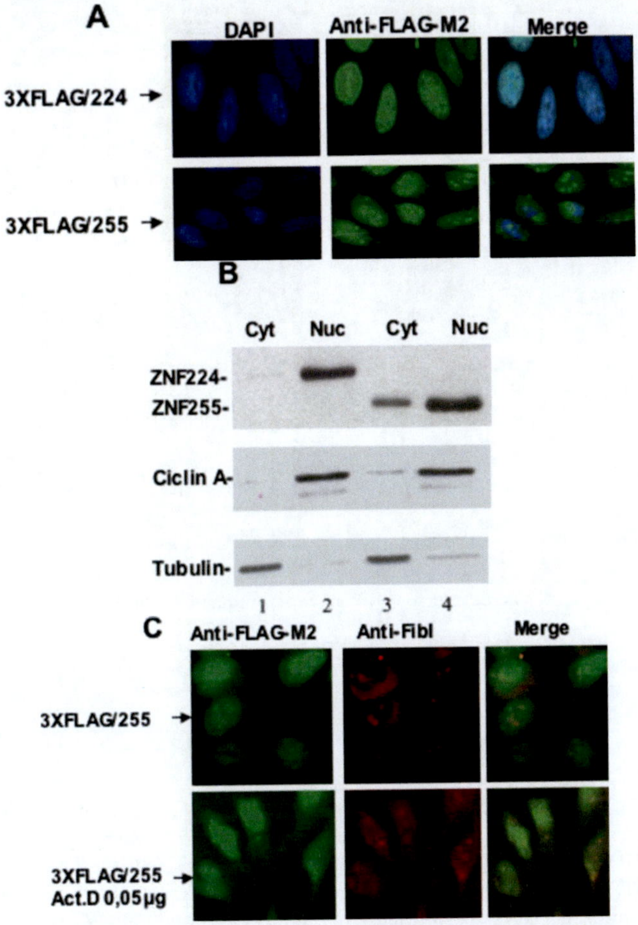


Figure 4. Cellular localization of the ZNF224 and ZNF255 proteins. (A) HeLa cells were transfected with two vectors expressing 3XFLAG-ZNF224 and 3XFLAG-ZNF255 fusion proteins, respectively. Cells were fixed and immunofluorescent microscopy was performed using anti-FLAG-M2 primary antibody and anti-mouse FITC secondary antibody (green). Cell nuclei were stained with DAPI. (B) Western blotting analysis was carried out to verify the expression of 3XFLAG-ZNF224 and 3XFLAG-ZNF255 fusion proteins through the use of anti-FLAG-M2 antibody. Lanes 1 and 3 show the data obtained with cytoplasmic extracts (Cyt), lanes 2 and 4 show the results obtained with nuclear extracts (Nuc). (C) HeLa cells were transfected with 3XFLAG-ZNF255 expression vector and after 48 h, the cells were treated with 0.05 μg/ml actinomycin D for 4 h. The cells were observed by immunofluorescence microscopy using anti-FLAG-M2 antibody (for the FLAG-ZNF255 fusion protein), anti-FibI (for the co-localization assay with fibrillarin) and anti-mouse FITC (green) and anti-rabbit (red) secondary antibodies. Reproduced with permission from Medugno et al. Differential expression and cellular localization of ZNF224 and ZNF255, two isoforms of the Krüppel-like zinc-finger protein family. (2007) Gene 403, 125-131. Elsevier.

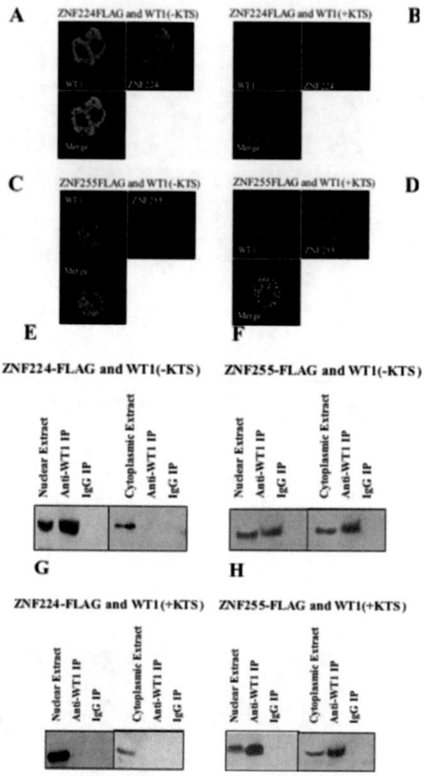


Figure 5. ZNF224 and ZNF255 colocalize with different WT1 isoforms. COS7 cells were transfected with p3XFLAG-ZNF224 plus pcDNA3-WT1(-KTS) (A) or pcDNA3-WT1(+KTS) (B) and with p3XFLAG-ZNF255 plus pcDNA3-WT1(-KTS) (C) or pcDNA3-WT1(+KTS) (D). Twenty-four hours after transfection, the cells were fixed, permeabilized and immunofluorescence was performed with anti –FLAG and anti-WT1(C-19) as primary antibodies. Panel A shows the overlap of ZNF224-FLAG (red) and WT1(-KTS) (green) signals, whereas ZNF224-FLAG does not overlap with the signal of WT1 (+KTS) (B). ZNF255-FLAG has a nuclear localization (red) that partially overlaps with WT1 (-KTS) signal (green) (C) but shows a nuclear speckled pattern that precisely overlaps with the signal of WT1 (+KTS) isoform (D). HEK293 cells were co-transfected with p3XFLAG-ZNF224 and pcDNA3-WT1 (-KTS) (E) or with p3XFLAG-ZNF224 and pcDNA3-WT1(+KTS) (G). C19 antibody was used to immunoprecipitate the different isoforms of WT1 from nuclear and cytoplasmic lysate. The immunoprecipitated complexes were probed with the anti-FLAG antibody. (E) shows a specific interaction of ZNF224 with WT1(-KTS) in the nucleus. No interaction was observed between ZNF224 and WT1(+KTS) (G). The same experiment was performed on HEK293 transfected with p3XFLAG-ZNF255 and pcDNA3-WT1(-KTS) (F) or pcDNA3-WT1(+KTS) (H). Western blotting shows that ZNF255 interacts with both WT1 isoforms in nuclear and cytoplasmic compartments. Reproduced with permission from Florio et al. Biochemical and functional interaction between ZNF224 and ZNF255, two members of the Kruppel-like zinc-finger protein family and WT1 protein isoforms. (2010) Hum Mol. Genet. 19, 3544–56. Oxford University Press.

## The ZNF224 Protein: A Multifaceted Player in the Nuclear Function

Despite the large number of KRAB-ZFPs that have been so far identified, several aspects of their biological role still remain elusive and need to be elucidated. It is known that some members of the KRAB-ZFP family are involved in the maintenance of the nucleolus, cell differentiation, cell proliferation, apoptosis and neoplastic transformation [2]. A down-regulation of some KRAB-ZFPs in various cancer cell types has been demonstrated to correlate with sustained proliferation and cell survival, thus suggesting a functional role of tumor suppressor gene for these proteins. Conversely, an over-expression of these KRAB-ZFPs results in an induction of apoptosis in tumor cells [39, 40]. To evaluate the possible involvement of ZNF224 in controlling cell proliferation, it was ectopically expressed in HeLa cells, using the Tet-off transgene regulation system to manipulate its expression.

Stably transfected cells, which express the ZNF224 protein depending on the concentration of doxycycline in the culture media, were analyzed for their ability to survive in a time-course experiment. Figure 6A shows that the induction of the ZNF224 expression, after the withdrawal of doxycycline from the culture media, inhibits the proliferation rate of the cells, suggesting that this KRAB-ZFP can act as an antiproliferative agent.

The ectopic expression of ZNF224 in HeLa cells up to 24 hours is responsible for a significant increase in the hypodiploid sub-G1 cell population (Figures 6B and C). Both results strengthen the argument that ZNF224 inhibits cell proliferation and then induces cell death.

In accordance with this finding, we propose that ZNF224 plays a role in the apoptosis pathway also through the interaction with WT1. We demonstrated that ZNF224 acts as a co-regulator of WT1 in the transcriptional modulation of apoptotic genes according to the cellular or tissue milieu in which WT1 is expressed. Indeed, by siRNA-mediated knockdown and over-expression of ZNF224 in erythroleukemia K562 cells, it was demonstrated that, acting as a co-activator of WT1, ZNF224 positively affects the normal physiological expression of the VDR gene and of other pro-apoptotic WT1 target genes, such as *bak* and *bax* [38, 41 and unpublished data]. Conversely, acting as a co-repressor of WT1, ZNF224 is able to prevent the transcriptional activation of anti-apoptotic WT1 target genes, such as *A1BFL1* and *bag3* [42, 43 and unpublished data]. These data indicate that, in addition to its function of transcriptional repressor, ZNF224 also functions as a cofactor of the WT1 transcription factor, and suggest that ZNF224/WT1 interaction plays a substantial role in leukemia cells (Figure 7).

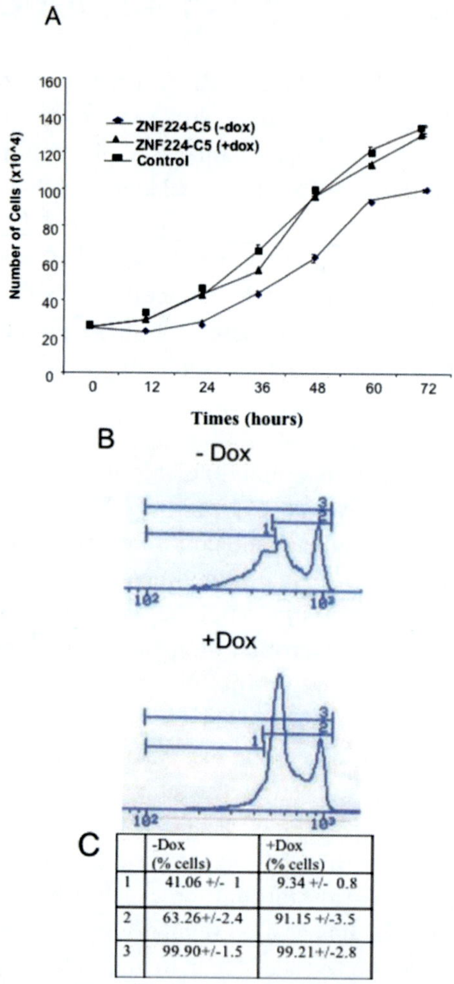


C

| | -Dox (% cells) | +Dox (% cells) |
|---|---|---|
| 1 | 41.06 +/- 1 | 9.34 +/- 0.8 |
| 2 | 63.26+/-2.4 | 91.15 +/-3.5 |
| 3 | 99.90+/-1.5 | 99.21+/-2.8 |

Figure 6. Effects of ZNF224 overexpression on the proliferation of HeLa cells. (A) Stable transfected cells overexpressing ZNF224 were cultured in presence (+Dox) and absence (-Dox) of 1.5 μg of doxycycline up to 72 hours and the percentage of proliferating and surviving cells was evaluated at various times. The removal of doxycycline (-Dox) causes a sustained expression of ZNF224 and a reduced cell proliferation compared with the cells that do not overexpress ZNF224 ( +Dox) and with the untransfected HeLa cells (control). (B) Flow cytometric analysis, after PI staining, of cells overexpressing ZNF224 (-Dox) and cells not overexpressing ZNF224 (+Dox). Cells with sub-G1 DNA content are shown. (C) In the table are the results of the experiment shown in (B). The percentages of the cells overexpressing ZNF224 (-Dox) and not overexpressing ZNF224 (+Dox) are indicated as numerical values.

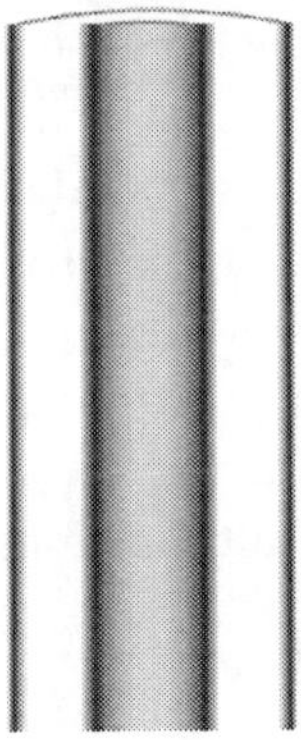

Figure 7. Proposed biological roles of ZNF224 and ZNF255. The interaction of the two isoforms with various molecular partners (WT1+KTS, WT1-KTS, DEPDC1, PRMT5) in different cellular contexts defines the broad spectrum of functions of the multifaceted ZNF224 protein.

Recently, the testis cancer antigen DEPDC1, involved in bladder carcinogenesis, was indicated as a molecular partner of ZNF224 (44). This study demonstrated that DEPDC1 acts as a transcriptional co-repressor of ZNF224 on the A20 gene, a negative regulator in the NF-kB antiapoptotic pathway. Therefore, A20 transcriptional repression by the DEPDC1-ZNF224 complex results in the suppression of apoptosis through the activation of the NF-kB signaling pathway. These data suggest that in bladder cancer cells, contrary to what is observed in leukemia cells, ZNF224 supports an anti-apoptotic function (Figure 7). In conclusion, the study of the multifunctional protein ZNF224 confirms that the ability of ZFPs to interact with other proteins in different cellular contexts may determine the broad spectrum of functions potentially attributed to this transcriptional regulator family.

# Conclusion

The KRAB-containing zinc finger proteins (KRAB-ZFPs) are the largest group of transcriptional repressors encoded in the genomes of mammals and, particularly, of vertebrate tetrapodes. In spite of their numerical abundance, the physiological functions of KRAB-ZFPs remain poorly defined and few of their target genes have been found. Therefore, ZNF224, the transcriptional repressor of the human aldolase A gene, is a useful molecular tool for studying the biological functions of this protein family.

ZNF224 and its alternatively spliced isoform, ZNF255, originating from a single gene, show different expression patterns in human tissues, a distinct subcellular localization, a specialized ability to interact with other protein partners and, consequently, functional differences. Furthermore, ZNF224 interacts with different partners, and through these interactions modulates the expression of genes involved in controlling the energy metabolism, cancer progression and apoptosis. Further studies will help to elucidate the full spectrum of the activities of the multifunctional protein ZNF224 and its role in various biological process.

# References

[1] Messina, D.N., Glasscock, J., Gish, W. and Lovett, M. An ORFeome-based analysis of human transcription factor genes and the construction of a microarray to interrogate their expression. (2004) *Genome Res.* 14, 2041-2047.

[2] Urrutia R. KRAB-containing zinc-finger repressor proteins. (2003) *Genome Biol.* 4, 231.

[3] Bellefroid, E.J., Poncelet, D.A., Lecocq, P.J., Revelant, O., and Martial, J.A. The evolutionarily conserved Kruppel-associated box domain defines a subfamily of ukaryotic multifingered proteins. (1991) *Proc. Natl. Acad. Sci. USA* 88, 3608-3612.

[4] Rosati, M., Marino, M., Franzè, A., Tramontano, A. and Grimaldi, G. Members of the zinc finger protein gene family sharing a conserved N-terminal module. (1991) *Nucleic Acids Res.*. 25, 5661-5667.

[5] Collins T., Stone, J.R. and Williams, A.J. All in the family: the BTB/POZ, KRAB, and SCAN domains. (2001) *Mol. Cell Biol.* 21, 3609-3615.

[6] Peng H., Begg, G.E., Harper, S.L., Friedman, J.R., Speicher, D.W. and Rauscher F.J. III. Biochemical analysis of the Kruppel-associated box (KRAB) transcriptional repression domain. (2000) *J. Biol. Chem.* 275, 18000-18010.

[7] Moosmann P., Georgiev, O., Le Douarin, B., Bourquin, J.P. and Schaffner W. Transcriptional repression by RING finger protein TIF1 beta that interacts with the KRAB repressor domain of KOX1. (1996) *Nucleic Acids Res.* 24, 4859-4867.

[8] Kim S.S., Chen, Y.M., O'Leary, E., Witzgall, R., Vidal, M. and Bonventre J.V. A novel member of the RING finger family, KRIP-1, associates with the KRAB-A transcriptional repressor domain of zinc finger proteins. (1996) *Proc. Natl. Acad. Sci. USA*, 93, 15299-15304.

[9] Lechner, M.S., Begg, G.E., Speicher, D.W., and Rauscher F.J. III. Molecular determinants for targeting heterochromatin protein 1-mediated gene silencing: direct chromoshadow domain-KAP-1 corepressor interaction is essential. (2000) *Mol. Cell Biol.* 20, 6449-6465.

[10] Schultz D.C., Friedman J.R., Rauscher F.J. III. Targeting histone deacetylase complexes via KRAB-zinc finger proteins: the PHD and bromodomains of KAP1 form a kooperative unit that recruits a novel isoform of the Mi-2alpha subunit of NurD. (2001) *Genes and Dev.* 15, 428-443.

[11] Schultz D.C., Ayyanathan, K., Negorev, D., Maul, G.G., and Rauscher F.J. III. SETDB1: a novel KAP1-associated histone H3, lysine 9-specific methyltransferase that contributes to HP1-mediated silencing of euchromatic genes by KRAB zinc finger proteins. (2002) *Genes and Dev.* 16, 919-932.

[12] Sripathy, S.P., Stevens, J. and Schultz, D.C. The KAP1 corepressor functions to coordinate the assembly of de novo HP1 demarcated microenviroments of heterochromatin required for Krab zinc finger protein mediated transcriptional repression. (2006) *Mol. Cell Biol.* 26, 8623-8638.

[13] Shannon, M., Hamilton, A.T., Gordon, L., Branscomb, E. and Stubbs, L. Differential expansion of zinc finger transcription factor loci in homologous human and mouse gene clusters. (2003) *Genome Res.* 13, 1097-1110.

[14] Huntley, S., Baggott, D.M., Hamilton, A.T., Tran-Gyamfi, M., Yang, S., Kim, J., Gordon, L., Bramscomb, E. and Stubbs, L. A comprehensive catalog of human KRAB-associated zinc finger genes: insights into the evolutionary history of a large family of transcriptional repressors. (2006) *Genome Res.* 16, 669–677.

[15] Tadepally, H.D., Burger, G. and Aubry, M. Evolution of C2H2 zinc finger genes and subfamilies in mammals: species-specific duplication and loss of clusters, genes and effector domains. (2008) *BMC Evolutionary Biology* 8, 176-192.

[16] Medugno, L., Costanzo, P., Lupo, A., Monti, M., Florio, F., Pucci, P. and Izzo, P. A novel zinc finger transcriptional repressor, ZNF224,

interacts with the negative regulatory element (AldA-NRE) and inhibits gene expression. (2003) *FEBS Letters* 534, 93-100.

[17] Han, Z.G., Zhang, Q.H., Ye, M., Kan, L.X., Gu, B.W., He, K.L., Shi, S.L., Zhou, J., Fu, G., Mao, M., Chen, S.J., Yu, L., and Chen, Z. Molecular cloning of six novel Kruppel-like zinc finger genes from hematopoietic cells and identification of a novel transregulatory domain KRNB. (1999) *J. Biol. Chem.* 274, 35741-35748.

[18] Lee, T.H., Lwu, S., Kim, J., and Pelletier, J. Inhibition of Wilms tumor 1 transactivation by bone marrow zinc finger 2, a novel transcriptional repressor. (2002) *J. Biol. Chem.* 277, 44826-44837.

[19] Medugno, L., Florio, F., Cesaro, E., Grosso, M., Lupo, A., Izzo,P. and Costanzo, P. Differential expression and cellular localization of ZNF224 and ZNF255, two isoforms of the Krüppel-like zinc-finger protein family. (2007) *Gene* 403, 125-131.

[20] Mark, C., Abrink, M. and Hellman, L. Comparative analysis of KRAB zinc finger proteins in rodents and in man: evidence for several evolutionarily distinct subfamilies of KRAB zinc finger genes. (1999) *DNA Cell Biol.* 18, 381-396.

[21] Rosati, M., Franz, A., Matarazzo, M.R., and Grimaldi, G. Coding region intron/exon organization, alternative splicing, an X-chromosome inactivation of the KRAB/FPB-domain-containing human zinc finger gene ZNF41. (1999) *Cytogenet. Cell Genet.* 85, 291-296.

[22] Hoovers, J.M., Mannens, M., John, R., Bliek, J., van Heyningen, V., Porteous, D.J., Leschot, N.J., and Little, P.F. High-resolution localization of 69 potential human zinc finger protein genes: A number are clustered. (1992) *Genomics* 12, 254-263.

[23] Young, J.M. and Trask B.J. The sense of smell: genomics of vertebrate odorant receptors. *Hum. Mol Genet.* 11, 1153-1160.

[24] Costanzo, P., Lupo, A., Medugno, L., D'Agostino, P., Zevino, C. and Izzo, P. PKC-dependent phosphorylation of the p97 repressor regulates the transcription of aldolase A L-type promoter. (1999) *FEBS Letters* 454, 61-66.

[25] Medugno, L., Florio, F., De Cegli, R., Grosso, M., Lupo, A., Costanzo, P., and Izzo, P. The Kruppel-like zinc finger protein ZNF224 represses aldolase A gene transcription by interacting with the KAP-1 corepressor protein. (2005) *Gene*,359:35–43.

[26] Cesaro, E., De Cegli, R., Medugno, L., Florio, F., Grosso, M., Lupo, A., Izzo, P. and Costanzo, P. The Kruppel-like zinc finger protein ZNF224 recruits the arginine methyltransferase PRMT5 on the transcriptional

repressor complex of the aldolase A gene. (2009) *J. Biol. Chem.* 284:32230–321.

[27] Fabbrizio, E., El Messaoudi, S., Polanowska, J., Paul, C., Cook, J.R., Lee, K.H., Negre, V., Rousset, M., Pestka, S., Le Cam, A. and Sardet, C. Negative regulation of transcription by the type II arginine methyltransferase PRMT5. (2002) *EMBO Rep.* 3, 641-645.

[28] Pal, S., Vishwanath, S.N., Erdjument-Bromage, H., Tempst, P. and Saif, S. Human SWI/SNF-Associated PRMT5 methylates histone H3 arginine 8 and negatively regulates expression of ST7 and NM23 tumor suppressor genes. (2004) *Mol. Cell. Biol.* 24, 9630-9645.

[29] Sleeman, J.E. and Lamond, A.I. Newly assembled snRNPs associate with coiled bodies before speckles, suggesting a nuclear snRNP maturation pathway. (1999) *Curr. Biol.* 9: 1065–1074.

[30] Zhang, S., Hemmerich, P. and Grosse, F. Nucleolar localization of the human telomeric repeat binding factor 2 (TRF2). (2004) *J. Cell Sci.* 117, 3935-3945.

[31] Huang,Z., Philippin, B., O'Leary, E., Bonventre, J. V., Kriz, W., and Witzgall, R. Expression of the transcriptional repressor protein Kid-1 leads to the disintegration of the nucleolus. (1999) *J. Biol Chem.* 274, 7640-7648.

[32] Payen, E., Verkerk, T., Michalovich, D., Dreyer, S. D., Winterpacht, A., Lee, B., De Zeeuw, C. I., Grosveld, F.., and Galiart, N. The centromeric/nucleolar chromatin protein ZFP-37 may function to specify neuronal nuclear domains. (1998) *J. Biol. Chem.* 273:9099-9109.

[33] Yano K., Ueki N., Oda T., Seki N., Masuho Y., Muramatsu M. Identification and characterization of human ZNF274 cDNA, which encodes a novel Kruppel-type Zinc-Finger Protein Having nucleolar targeting ability. (2000) *Genomics* 65, 75-80.

[34] Rivera, M.N., and Haber, D.A. Wilms' tumour: connecting tumorigenesis and organ development in the kidney. (2005) *Nat. Rev. Cancer* 5, 699–712.

[35] Hohenstein, P., and Hastie, N.D. (2006). The many facets of the Wilms' tumour gene, WT1. Hum. Mol. Genet. 15 (Spec No 2), R196–R201.

[36] Yang L, Han Y, Suarez Saiz F, Minden MD. A tumor suppressor and oncogene: the WT1 story. (2007) *Leukemia* 21: 868-876.

[37] Hammes, A., Guo, J.K., Lutsch, G., Leheste, J.R., Landrock, D., Ziegler, U., Gubler, M.C., Schedl, A. Two splice variants of the Wilms' Tumor 1 gene have distinct functions during sex determination and nephron formation. (2001) *Cell* 106, 319-329.

[38] Florio F, Cesaro E, Montano G, Izzo P, Miles C, Costanzo P. Biochemical and functional interaction between ZNF224 and ZNF255, two members of the Kruppel-like zinc-finger protein family and WT1 protein isoforms. (2010) *Hum. Mol. Genet.* 19, 3544–56.

[39] Huang, C., Yang, S., Ge, R., Sun,H., Shen, F., Wang, Y., ZNF23 induces apoptosis in human ovarian cancer cells. (2008) *Cancer Lett.* 266, 135-143.

[40] Cheng, Y., Geng,H., Cheng, S.H., Liang, P., Bai, Y., Srivastava, G., Ng, M.H., Fukugawa, T., Wu, X., Chan, A.T., Tao, Q. KRAB zinc finger protein ZNF382 is a proapoptotic tumor suppressor that represses multiple oncogenes and is commonly silenced in multiple carcinomas. (2010) *Cancer Res.* 70, 6516-6526.

[41] Morrison, D.J., English, M.A. and Licht, J.D. WT1 induces apoptosis through transcriptional regulation of the proapoptotic Bcl-2 family member Bak. (2005) *Cancer Res.* 65, 8174–8182.

[42] Simpson L.A. Burwell E.A. Thompson K.A. Shahnaz S. Chen A.R. Loeb D.M. The antiapoptotic gene A1/BFL1 is a WT1 target gene that mediates granulocytic differentiation and resistance to chemotherapy. (2006) *Blood* 107,4695–4702.

[43] Cesaro E, Montano G, Rosati A, Crescitelli R, Izzo P, Turco MC, Costanzo P. WT1 protein is a transcriptional activator of the antiapoptotic bag3 gene. (2010) *Leukemia* 24, 1204-6.

[44] Harada Y, Kanehira M, Fujisawa Y, Takata R, Shuin T, MikiT, FujiokaT, Nakamura Y, Katagiri T. Cell-permeable peptide DEPDC1-ZNF224 interferes with transcriptional repression and oncogenicity in bladder cancer cells. (2010) *Cancer Res.* 70, 5829-39.

In In: Zinc Fingers
Editors: R. Ciofani and L. Makrlik
ISBN 978-1-62100-230-7


*Chapter 5*

# Diverse Roles of CCCH Zinc Finger Proteins in Stress and Plant Development Responses

***Balasubrahmanyam Addepalli*[1] *and Arthur G. Hunt*[2]**
[1]Rieveschl Laboratories for Mass Spectrometry,
Department of Chemistry, University of Cincinnati, OH 45221
[2]Department of Plant and Soil Sciences, University of Kentucky,
Lexington, KY 40546

## Abstract

Proteins containing zinc finger domains are highly abundant in eukaryotes. Of these, the CCCH class of motifs, where a zinc ion is postulated to be coordinated by three cysteine and one histidine residues, are present in a large number of proteins. Recent studies have implicated the members of this class in a wide variety of cellular functions ranging from developmental activities to amelioration of biotic and abiotic stresses. Animal CCCH zinc finger proteins recognize destabilizing AU-rich elements in the 3'-untranslated regions of messenger RNAs targeting them for degradation. Genetic analysis of plant proteins indicates their involvement in mRNA processing, floral morphogenesis, senescence, hormone-mediated environmental, biotic and abiotic stress responses. Furthermore, plant proteins exhibit differential expression patterns indicating diverse biological roles. Several plant proteins have similar functions and properties, such as trafficking between the nucleus and cytoplasm, and co-localizing with markers of processing bodies (PBs) or stress granules, although such proteins differ in zinc finger motif number they possess. Biochemical studies reveal that individual motifs of tandem fingers might participate in protein-protein interactions, nucleic acid binding, contributing ribonuclease activity or even calmodulin-mediated calcium signaling. This chapter summarizes the recent advances and current thinking of this protein family in plants including genome level characterization, classification, nucleic acid binding features, site directed mutagenesis based functional characterization, and finally, the future directions to fill the gaps in structural information.

## Introduction

Proteins in multicellular organisms contain self-contained autonomously folded regions of amino acid sequences called domains or modules with distinctive secondary structure content and function (Campbell and Downing, 1994). Such regions may also exhibit consensus sequences that may be identified without bioinformatic analysis. Often those sequences, referred to as motifs, are shared between distinct proteins with different biological function and even from diverse phyla (Aitken, 1999). They include extensive range of phosphorylation site motifs, metal binding sites, nucleotide binding and covalent attachment sites. Of all the metal binding motifs observed in proteins, zinc coordinating amino acid ligands exhibit highest diversity. In a zinc motif, conserved cysteines alone or in combination with histidines serve as ligands to a zinc ion whose coordination is essential to stabilize the tertiary structure of the protein. These structures may be present singularly or in tandem in a given protein.

The classical zinc fingers are DNA binding motifs observed in transcription factors (Takatsuji, 1998; Moore and Ullman, 2003), but now these are known to bind RNA (Brown, 2005) and mediate protein-protein interactions (Gamsjaeger *et al.*, 2007). One such class of zinc finger motif is the CCCH motif with a canonical sequence C-X-C-X-C-X-H (C-$X_{6\text{-}14}$-C-$X_{4\text{-}5}$-C-$X_3$-H, X being any other amino acid) (Berg and Shi, 1996). Members of this class include the factor that recognizes mRNA-destabilizing elements (tristetraprolin, or TTP) (Blackshear, 2002), and the 30-kDa subunit of the mammalian Cleavage and Polyadenylation Specificity Factor (CPSF30) and its yeast counterpart Yth1 (Barabino *et al.*, 1997). The CCCH motif was originally identified in TTP and binds to AU-rich elements (ARE) in mRNA (Carrick *et al.*, 2004; Lai *et al.*, 1999). In plants, the family of CCCH motif containing proteins is large with 68 and 67 members in *Arabidopsis* and rice, respectively. This review focuses on current advances in the knowledge base of plant family of CCCH proteins.

## Classification of Plant CCCH Proteins

Wang *et al.* (2008) performed detailed phylogenetic analysis of *Arabidopsis* and rice CCCH proteins. They observed that plant members exhibit high diversity for motif number and spacing amino acids between adjacent fingers in each protein. Based on the alignment of full-length CCCH motif containing-sequences (a total of 68) and bootstrap values (>90%) in the outer clades of aligned tree, the *Arabidopsis* CCCH family is divided into 11 subfamilies (designated I to XI) with 1 to 6 motifs in a given protein. Individual members within a subfamily displayed structural similarities in terms of the number of conserved tandem motifs, intron/exon structure, and presence of other domains (like WD40, RRM or KH motifs) (Wang *et al.*, 2008). Excepting the members of subfamily IX, almost all other subfamily members of *Arabidopsis* have homologues in other plants including moss. In other words, the subfamily IX is restricted to the vascular plants.

Further, the proteins with sequence signatures of $C\text{-}X_{7\text{-}8}\text{-}C\text{-}X_5\text{-}C\text{-}X_3\text{-}H$ predominate in both *Arabidopsis* and rice suggesting a common ancestor for these proteins. Web Logo analysis of these proteins showed that >75% of CCCH motifs are rich in glycine and phenylalanine, prompting the revision of signature sequence as one to six $C\text{-}X_{4\text{-}15}\text{-}C\text{-}X_{4\text{-}6}\text{-}C\text{-}X_3\text{-}H$ motifs that are rich in glycine and phenylalanine residues (Figure 1). More than 50% of CCCH genes in rice and *Arabidopsis* are expressed in all tissues, 25% of them are not found to be expressed under tested conditions, and the remaining showed tissue-specific expression thus revealing diverse expression patterns in the family. The vascular plant specific subfamily IX possesses a highly conserved $C\text{-}X_5\text{-}C\text{-}X_4\text{-}C\text{-}X_3\text{-}H$ motif along with ankyrin repeats (ANK) the motif that mediates protein-protein interactions. These genes seem to be activated or suppressed in response to biotic or abiotic stresses.

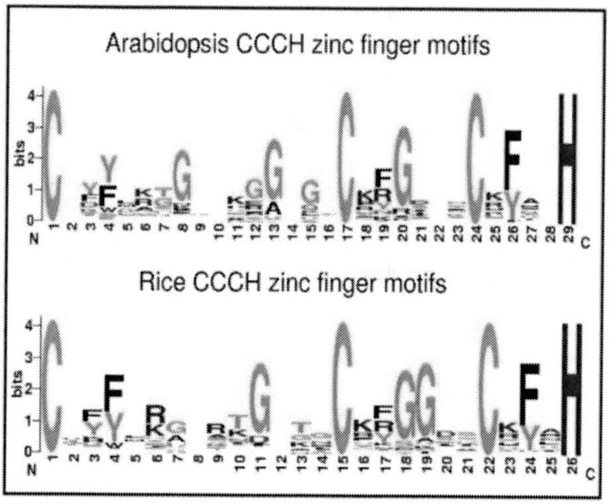


Figure 1. Sequence logos for the CCCH zinc finger motifs of *Arabidopsis* and rice. Numbers on the x-axis represent the sequence positions in zinc finger motifs. The y-axis represents the information content measured in bits. Reproduced from original publication by Wang *et al. BMC Genomics* 2008 9:44/doi:10.1186/1471-2164-9-44.

## CCCH-Motif Structure, Nucleic Acid Binding and RNA Processing Activities

Most of the CCCH proteins, when expressed in bacteria with small peptide affinity tags such as the His tag, fall out of solution during purification (Blackshear *et al.*, 2005). However, several *Arabidopsis* CCCH proteins could be expressed and purified as maltose binding protein fusions with no insolubility issues (Addepalli and Hunt, 2008), although cleavage of fusion tag leads to precipitation of tag free protein (our unpublished observations) thus precluding the ability for structural characterization.

Nevertheless, a handful of solution structures of tandem zinc finger motifs are available for animal proteins including TTP (Hudson *et al.*, 2004) in complex with RNA nonamer 5'-UUAUUUAUU-3' (PDB ID: 1RGO), CCCH motifs 2 and 3 of the human CPSF30 (PDB ID: 2D9N), and the muscleblind-like (MBNL) protein (PDB ID: 3D2S). The first and second CCCH motifs of the *Arabidopsis* AtCPSF30 can be subjected to homology modeling with human CPSF30 CCCH motif 2 and 3 (2D9N) with a reasonable fit (Figure 2).

The mechanism of RNA AU element (5'-UAUUUAUU-3') recognition and binding by animal CCCH zinc finger protein tristeraprolin has been well documented (Lai *et al.*, 2002; Hudson *et al.*, 2004). The binding interface between protein and RNA involve hydrophobic packing and hydrogen-bonding with prominent stacking interactions between the conserved aromatic amino acids Tyr170, Phe176, Tyr208, Phe214, and all of the RNA bases (Hudson *et al.*, 2004). Substitution of any one of these aromatic amino acids abrogated the RNA binding (Lai *et al.*, 2002).

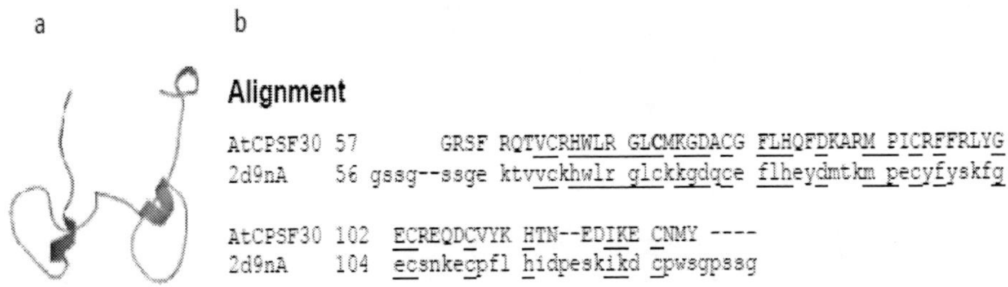


Figure 2. Structural model (a), and sequence alignment of AtCPSF30 (b) with human CPSF30 CCCH motif 2 and 3. An automated comparative protein modeling of AtCPSF30 sequence with known zinc finger structures on the SWISS-MODEL workspace (http://swissmodel.expasy.org /workspace/index.php?func=show_workspace) (Guex and Peitsch, 1997; Schwede *et al.*, 2003; Arnold *et al.*, 2006) has selected only the first two CCCH motifs with amino acid sequence from 57-124 for alignment with the NMR solution structure of human CPSF30 CCCH motifs 2 and 3.

In the model of AtCPSF30, candidate side chains for analogous stacking interactions can be identified in the first (RNA-binding) CCCH motif; however, site-directed mutagenesis of Phe61 or Trp69 residues to alanine did not disrupt RNA binding (our unpublished observations), suggesting an involvement of other side chains, and perhaps mechanisms other than base-aromatic side chain stacking. The sequence homology of AtCPSF30 with the animal counterparts is limited to the zinc finger region; however, the N-terminal and C-terminal regions of AtCPSF30 exhibit high sequence similarity with the corresponding regions of other plant homologues indicating the unique functional nature of these plant specific motifs (Figure 3).

Nevertheless, AtCPSF30 has been shown to be a legitimate polyadenylation factor because of its domain specific interactions with a number of polyadenylation factor subunits (Delaney *et al.*, 2006; Hunt *et al.*, 2008; Rao *et al.*, 2009), summarized in Figure 3. For example, AtFip1(v) regulates the endonuclease activity of AtCPSF30 through protein-protein interactions (Addepalli and Hunt, 2007) involving the third CCCH motif, thus establishing that AtCPSF30 is a processing endonuclease in the plant polyadenylation apparatus.

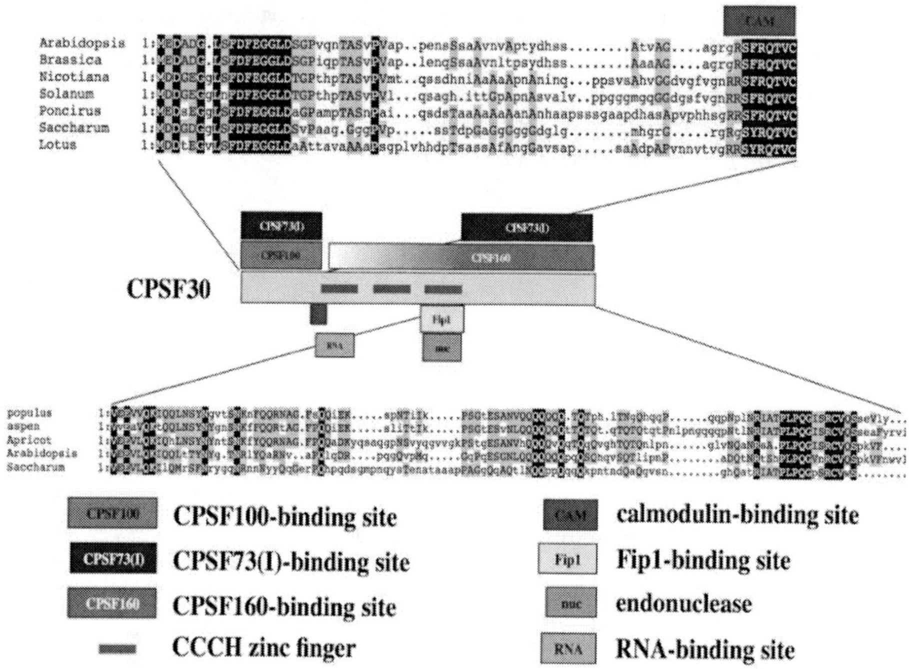


Figure 3. Functional map of the *Arabidopsis* CPSF30 ortholog. A one-dimensional representation of the amino acid sequence (light blue box) is shown in the center of the figure. The color-coding of the various motifs is described at the bottom of the figure. Amino acid sequence alignments of the plant-specific N-terminal and C-terminal domains are shown above and below the functional map of AtCPSF30, respectively. Accession numbers that correspond to the N-terminal sequences: Brassica – CD828769; Nicotiana – CK286112; Solanum – CO502465; Poncirus – CV705634; cane – CA096547; Lotus – BP037936. Accession numbers that correspond to the C-terminal sequences: populus – CV243319; aspen – CK088987; Apricot – CV053103; *Arabidopsis* – AY140901. Reproduced from original publication by Rao *et al. BMC Cell Biology* 2009 10:51/ doi:10.1186/1471-2121-10-51.

Other plant CCCH motif proteins that have been characterized for RNA binding activities include HUA1 (At3g12680) (Li *et al.*, 2001), AtSmic1 (At1g21570), At1g04990, At3g51950, At2g05160 and At5g56930 (Addepalli and Hunt, 2008), and AtZF1 (At2g25900) (Pomeranz *et al.*, 2010). Both HUA1 and AtZF1 bind DNA and RNA under *in vitro* conditions. OsLIC (Os06g07043) has affinity for DNA and polyA, polyU and polyG, but not polyC (Wang *et al.*, 2008). AtCPSF30, AtSmic1, At1g04990, At3g51950,

At2g05160 and At5g56930 all possess endonuclease activity (Addepalli and Hunt, 2008).

Interestingly, with AtCPSF30 and AtSmic1, the RNA binding and ribonuclease activities map to individual CCCH motifs, as specific mutations in a given CCCH motif affected either RNA binding or nuclease activity but not both (Addepalli and Hunt, 2007, 2008). However, other plant proteins exhibiting ribonuclease activity have not yet been associated with specific cellular processes, so far.

## Oxidation States of Cysteine and Metal Binding Characteristics

Cysteine is the most redox-sensitive amino acid, although the sensitivity depends on the thiol-disulfide oxidation potentials of the accessible thiol groups which in turn determined by the chemical environment in protein (Gilbert, 2006).

Formation of a disulfide bond through oxidation of cysteine is the most common post-translational modification associated with protein folding. It is increasingly realized that changes in the redox state of cysteine residues or oxidation of specific cysteine residues is one way to modify the structure and thus the activity (Hogg, 2003) of proteins including kinases (Nakashima *et al.*, 2002), phosphatases (Cho *et al.*, 2004), and those involved in regulating the gene expression (Lee and Maret, 2001).

Bacterially expressed AtCPSF30 shows oxidized form of cysteines in the third CCCH motif thus forming a disulfide bond between Cys128 and Cys134 (Addepalli *et al.*, 2010). The endonuclease activity of AtCPSF30 requires the presence of a disulfide bond in the third CCCH motif (Addepalli and Hunt, 2008). Treatment of protein with sulfhydryl agents like DTT or 2-mercaptoethanol abolishes the nuclease activity (Addepalli and Hunt, 2008) with concomitant disappearance of disulfide bond (Addepalli *et al.*, 2010).

Interestingly, micromolar concentrations of zinc inhibited the endonuclease activity of AtCPSF30 suggesting rearrangement of protein conformation in the presence of excessive zinc (Addepalli and Hunt, 2008). From these considerations, it may be essential to investigate the redox state of cysteines in CCCH motifs by any one of many analytical methods (Camerini *et al.*, 2005); (Lee and Michel, 2010); (Addepalli and Hunt, 2008), especially if a known activity is attributed to a given CCCH motif.

The CCCH zinc finger of TTP can coordinate either zinc or iron, and this metal ion binding leads to correct protein folding and recognition of physiologically relevant RNA sequence with equal affinity (diTargiani *et al.*, 2006). Furthermore, when metal-coordinated TTP peptides are treated with hydrogen peroxide ($H_2O_2$), complete oxidation of cysteines leading to disulfide bond formation occurred with ferrous substituted CCCH-motif peptides in contrast to the 40% oxidation of zinc substituted version. This suggests that iron-containing motifs are more prone to oxidation, probably because of the generation of hydroxyl radicals from the reaction of ferrous-bound TTP with $H_2O_2$ via Fenton chemistry (Lee and Michel, 2010). Interestingly, the twin CCCH-motif construct of TTP (tristetraprolin-2D) exhibits visible red coloration when purified from *E. coli* unless it is expressed under iron-restricted conditions (diTargiani *et al.*, 2006). It has thus been proposed that both zinc and iron-coordinated CCCH motifs may possibly be present in the physiological conditions and there could be a link between iron and oxidative stress for TTP (Lee and Michel, 2010). Interestingly, several plant CCCH proteins including AtCPSF30 are also dark red when expressed and purified from *E. coli* (our unpublished observations). Metal analysis of overexpressed MBP-AtCPSF30 fusion protein by Inductively Coupled Plasma Mass Spectrometry (ICP-MS) revealed higher amount of iron compared to zinc (B Addepalli, unpublished data). More research is needed to verify whether the postulated link between iron and oxidative stress observed in case of TTP also be occurring in plants. In addition, X-ray crystal structures may shed more light on types of metal ion coordination observed in these proteins in comparison to the inability of NMR solution structures to deliver such information.

## Role of Plant CCCH Proteins in Stress Responses

A number of CCCH-type zinc finger proteins including AtCPSF30 (Zhang *et al.*, 2008) and GhZFP1 (*Gossypium hirsutum*) (Guo *et al.*, 2009) have been implicated in abiotic and biotic stress responses. These stresses cause an increase in levels of reactive oxygen species (ROS) at certain stage of stress exposure. They also cause signal-specific changes in intracellular $Ca^{2+}$ levels, which along with calmodulin regulate diverse physiological processes that are important for stress adaptation (Reddy *et al.*, 2011). For example, the $H_2O_2$

detoxifying enzyme catalase was activated by calmodulin suggesting a calcium-mediated signaling as an important response to reactive oxygen species (ROS) (Yang and Poovaiah, 2002). AtCPSF30 possesses calmodulin recognition motif that abuts the first (RNA-binding) CCCH motif (Delaney *et al.*, 2006). Moreover, calmodulin inhibits RNA binding of the protein in a calcium-dependent fashion (Delaney *et al.*, 2006). Thus, RNA and calmodulin binding by AtCPSF30 are mutually exclusive because of the occupancy of same surface of AtCPSF30 by calmodulin and RNA, suggesting a link between RNA processing and stress responses. Consistent with this, mutant plants deficient in AtCPSF30 expression show an altered response to oxidative stress (Zhang *et al.*, 2008).

A twin CCCH zinc finger motif (Cx8Cx5Cx3H and Cx5Cx4Cx3H) containing-protein GhZFP1 was isolated by Guo *et al* (2009) using a differential hybridization screen of a salt-induced cotton (*G. hirsutum*) cDNA library (Guo *et al.*, 2009). The expression of *GhZFP1* gene was induced by salt, drought and salicylic acid. Overexpression of this gene resulted in enhanced salt stress and resistance to *Rhizoctonia solani.* Similarly, disruption of the expression of CCCH-motif encoding genes AtSZF1 and AtSZF2 lead to salt sensitive phenotype while over expression of AtSZF1 alone resulted in enhanced tolerance to salt stress (Sun *et al.*, 2007).

A similar salt-inducible CCCH protein (MsZFN) was isolated from *Medicago sativa*; this protein has a sugar transport protein sequence signature at the C-terminal end (Chao *et al.*, 2009). While GhZFP1, MsZFN, AtSZF1 and AtSZF2 are nuclear-localized proteins (Sun *et al.*, 2007; Chao *et al.*, 2009; Guo *et al.*, 2009), location of the AtCPSF30 is defined by its interactions with other CPSF subunits. AtCPSF30 also co-localizes with AtDcp2 in P-bodies, the locations of mRNA decapping and breakdown. This association of AtCPSF30 with AtDcp2 is interesting, as it insinuates an endonuclease into the mRNA-degrading system (Rao *et al.*, 2009) (Figure 4). Similarly, a tandem CCCH-motif subfamily IX member AtTZF1 localizes to the nucleus and to cytoplasmic P-bodies (Pomeranz *et al.*, 2010). This protein could bind to both DNA and RNA and is postulated to control gene expression through posttranscriptional regulation (see below). Another nuclear CCCH-type zinc finger protein in rice, OsDOS, has been implicated in delaying the leaf senescence by integrating developmental cues with the jasmonate pathway (Kong *et al.*, 2006).

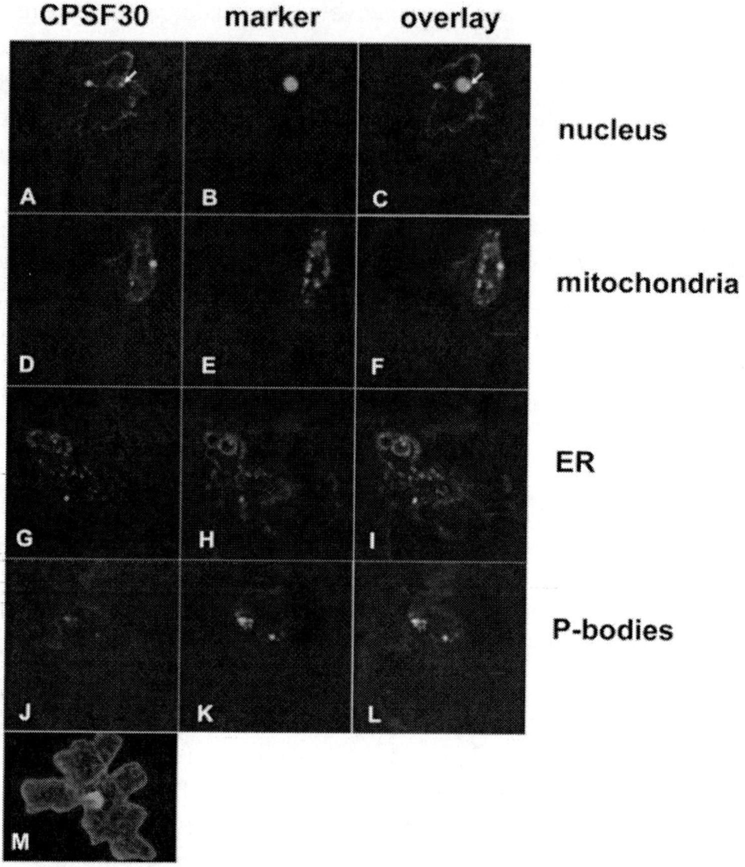


Figure 4. Subcellular distribution of AtCPSF30 in tobacco cells. In this figure, the overlays of the two images corresponding to pairs of fusion proteins are shown in the column on the right (panels C, F, I, and L). The AtCPSF30 is visualized in panels A, D, G, and J. Markers used to assess the subcellular location of AtCPSF30 were: nuclear marker (GFP-AtZFP11), panel B; mitochondrial marker (GFP-NDA2), panel E; endoplasmic reticulum marker, panel H; and the P-body marker (Dcp2), panel K. The distribution of unmodified DSR is shown in panel M. Red bars are 10 μm size markers. Reproduced from original publication by Rao *et al. BMC Cell Biology* 2009 10:51/ doi:10.1186/1471-2121-10-51.

## Roles of CCCH Proteins in Developmental Responses

The Arabidopsos CCCH–type zinc finger proteins HUA1 (Li *et al.*, 2001) and FES1 (Schmitz *et al.*, 2005) have been implicated in flower development.

AGAMOUS (AG), a key player in flower development specifies reproductive organ identities and regulates the timely transition of stem cells in the floral meristem. HUA1 in combination with HEN4 control the processing of AG pre-mRNA. The nucleus-localized HUA1 protein has six tandem CCCH-motifs that spans the C-terminal two-thirds of the protein, and binds preferentially to poly rU or poly rG homopolymers (Li *et al.*, 2001). HUA1 interacts with HEN4 (HUA ENHANCER4), a K homology (KH) domain-containing RNA binding protein, and promotes the processing of AGAMOUS (AG) pre-mRNA (Cheng *et al.*, 2003). Mutations in HUA1 or HEN4 compromised AG pre-mRNA processing leading to floral defects similar to *agamous* mutants including loss of reproductive-to-perianth organ transformation and loss of floral determinacy.

The suppression of flowering and maintenance of the winter-annual habit of *Arabidopsis* require FRIGIDA (FRI), FRIGIDA-ESSENTIAL 1 (FES1), and FRIGIDA-LIKE 1 (FRL1) which act in cooperative manner to promote upregulation of the flowering repressor FLOWERING LOCUS C (FLC). The *FES1* gene encodes a protein with a single CCCH-type zinc finger that promotes the expression of FLC in a FRI-dependent manner (Schmitz *et al.*, 2005). FES1 is expressed in the vascular system and shoot/root apex tissues, and is localized to nucleus. Beyond this, the biochemical mechanism of increase in FLC expression mediated by FRI, FES1 and FRL1, is unknown.

A number of CCCH-motif containing proteins have been associated with embryo development and seed germination in *Arabidopsis* and cucumber. The *Arabidopsis* proteins PEI1 (Li and Thomas, 1998) and its homologue CsSEF1 (Grabowska *et al.*, 2009), each with two CCCH motifs preceded by one CHCH-type zinc finger motif, are expressed throughout the *Arabidopsis* and cucumber embryos with maximal expression from the heart stage through the cotyledon stage. Because PEI1 binds to DNA, this protein is thought to be a transcription factor. Antisense mediated disruption of PEI1 gene resulted in interruption of embryogenesis at the heart stage leading to aberrant white seeds with enlarged heart-like embryo structures. Furthermore, several genes involved in chlorophyll synthesis were turned off in antisense-PEI1 seeds. Although the aberrant embryos could be rescued in embryo culture, they exhibited late germination and abnormal apical organs with cotyledon-like or leaf-like structures. However, the embryonic roots grew like normal roots and plants regenerated from this root callus turned green. Interestingly, the developing seed from these regenerated plants were exactly like anti-PEI1 seed (Li and Thomas, 1998). Transformation of *Arabidopsis* with antisense fragment of CsSEF1 led to abnormalities in zygotic embryos and cotyledons (Grabowska *et al.*, 2009).

## Role of CCCH –Motif Containing Proteins in Hormonal Responses

A few CCCH zinc finger proteins have been associated with hormone responses such as ABA, GA and brassinosteroids. In *Arabidopsis* seeds, PHYTOCHROME-INTERACTING FACTOR3-LIKE5 (PIL5) plays a key role in converting light signals to alterations in the ABA and GA pathways (reviewed in by (Oh *et al.*, 2007)). Mutant *pil5* plants exhibit elevated level of GA, decreased levels of ABA, and increased responses to GA irrespective of light conditions. In an attempt to identify the components involved in phytochrome and PIL5-mediated light signaling processes, a novel mutant called *somnus (som)* was isolated. This mutant germinates in complete darkness just like the *pil5* mutant. Loss of function of SOM caused early germination, increased GA accumulation and decreased ABA. Both PIL5 and SOM are nucleus-localized CCCH zinc finger proteins with two CCCH-motifs that were preceded by a CHCH-type motif in its N-terminus (Dong *et al.*, 2008). SOM is a seed specific gene with a progressively increasing gene expression during maturation but no expression in adult organs or tissues.

The phytochromes PHYA and PHYB repress the expression of SOM via PIL5 in imbibed seeds. Genetic analysis revealed that PIL5 positively regulates the expression of SOM by binding directly to its promoter. This data suggested that light regulates ABA and GA metabolic genes via PIL5 and then via SOM in seeds. However, the molecular function of CCCH zinc finger motif in SOM is unknown. It is not clear whether this protein acts as transcription factor regulating hormone metabolic genes or as an RNA binding protein, or a protein with some other function (Dong *et al.*, 2008).

The *Arabidopsis* tandem zinc finger protein AtTZF1 described above has recently been reported to play roles in hormone-mediated growth and environmental responses through regulation of gene expression (Lin *et al.*, 2011). This protein positively regulates ABA/sugar responses while negatively regulating the GA responses through modulation of gene expression. Like SOM, antisense-mediated RNA interference of *AtTZF1* caused early germination and slightly stress-sensitive phenotypes, whereas over-expressors of *AtTZF1* exhibited compact growth, late flowering and a stress-tolerant phenotype.

These phenotypic effects are observed to be because of compromised GA or ABA responses rather than due to changes in hormone levels. GA-Stimulated *Arabidopsis* (GASA) genes are a class of small, secreted protein molecules that play a major role in signal transduction pathways.

A gene network centered on one such GA-inducible but ABA/sugar-repressible peptide hormone encoded by *GASA6* is negatively regulated by *AtTZF1* over-expression. Hence AtTZF1 is postulated to be a regulator connecting sugar, ABA, GA and peptide hormone responses (Lin *et al.*, 2011).

One of the significant morphological markers for the brassinosteroid response in rice is leaf angle (Yamamuro *et al.*, 2000). Disruption of BR biosynthesis or its signal transduction pathway results in erect leaves. Treatment of rice seedlings with BRs result in increased leaf angle in a dose-dependent manner. OsLIC (*Oryza sativa* leaf and tiller angle increased controller) is a single CCCH-motif containing protein with a novel EELR domain, and its expression is significantly enhanced by 24-epibrassinolide, the most bioactive form of brassinosteroid in plants.

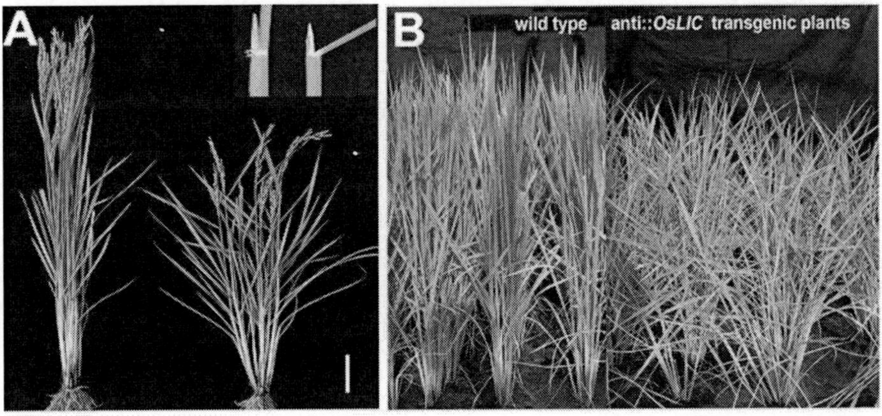


Figure 5. Phenotypes of antisense OsLIC transgenic plants. (A) Antisense OsLIC transgenic plants displayed multiple phenotypes, such as increased leaf angle, semi-dwarfism and increased tillering in mature period. Close-up view of lamina joint of wild-type plants and that of transgenic plants is shown in the right top corner of Figure 5A to display significantly increased leaf angle in transgenic plants. Bars = 10 cm. (B) T3 progeny of antisense OsLIC transgenic plants displayed similar phenotypes as T2 progeny in the field. Reproduced from the original publication by Wang, *et al* PLoS ONE 2008 3:e3521/ doi:10.1371/journal.pone. 0003521.g001.

This gene exhibited preponderant expression in the collar region, stem node and basal region of elongating internodes. Accordingly, down-regulation led to increased leaf angle, tiller angle, and reduced plant height (Figure 5).

The *AntiOsLIC* lines exhibited higher levels of sitosol and stigmasterol but its precursor 24-methylenechosterol remain unchanged compared to wild type (Wang *et al.*, 2008), indicating that OsLIC controls the rice architecture by regulating the sterol homeostasis in rice. The CCCH-motif of OsLIC protein could bind to dsDNA and ssRNA like polyA, polyU and polyG, indicating its dual specificity. However, the mechanism of gene regulation effected by this protein is not clear.

## Conclusion

The plant CCCH-motif containing zinc finger proteins are beginning to be characterized at the genetic, physiological and biochemical levels and their role in biological processes is slowly appreciated. Some of these proteins are postulated to be transcriptional activators based on their DNA binding ability or transcriptional activity in yeast. Although this could be true in some cases, this property is not applicable to all. For example, proteins like AtCPSF30 function as transcriptional activator in yeast although they bind exclusively to RNA. Hence the evidence of sequence specific DNA binding or promoter occupancy may be a more appropriate criterion to designate a given CCCH zinc finger as a transcriptional regulator. Moreover, beyond the nucleic acid binding specificities for DNA and/or RNA, biochemical functions of plant CCCH class of proteins are largely unknown. This type of information becomes essential to understand their biological function especially if these proteins are involved in control of metabolic, hormone or development responses through regulation of gene expression at posttranscriptional level. Because of the ability of CCCH-motifs to function as redox switches (as observed in case of TTP and AtCPSF30), the oxidation status of CCCH-motif/s need to be understood, especially when a protein is associated with a regulatory response. Such information coupled with structural data would help better understanding the role of CCCH class of proteins in growth and development of plants.

# Acknowledgment

Research in Arthur Hunt's laboratory was supported by NSF grants MCB-0313472 and IOS-0817818.


# References


Addepalli B, Hunt AG (2007) A novel endonuclease activity associated with the *Arabidopsis* ortholog of the 30-kDa subunit of cleavage and polyadenylation specificity factor. *Nucleic Acids Research* 35: 4453-4463.

Addepalli B, Hunt AG (2008) Redox and heavy metal effects on the biochemical activities of an *Arabidopsis* polyadenylation factor subunit. *Archives of Biochemistry and Biophysics* 473: 88-95.

Addepalli B, Hunt AG (2008) Ribonuclease activity is a common property of *Arabidopsis* CCCH-containing zinc-finger proteins. *FEBS Letters* 582: 2577-2582.

Addepalli B, Limbach PA, Hunt AG (2010) A disulfide linkage in a CCCH zinc finger motif of an *Arabidopsis* CPSF30 ortholog. *FEBS Letters* 584: 4408-4412.

Aitken A (1999) Protein consensus sequence motifs. *Molecular Biotechnology* 12: 241-253.

Arnold K, Bordoli L, Kopp Jr, Schwede T (2006) The SWISS-MODEL workspace: a web-based environment for protein structure homology modelling. *Bioinformatics* 22: 195-201.

Barabino SM, Hubner W, Jenny A, Minvielle-Sebastia L, Keller W (1997) The 30-kD subunit of mammalian cleavage and polyadenylation specificity factor and its yeast homolog are RNA-binding zinc finger proteins. *Genes and Development* 11: 1703-1716.

Berg JM, Shi Y (1996) The galvanization of biology: a growing appreciation for the roles of zinc. *Science* 271: 1081-1085.

Blackshear PJ (2002) Tristetraprolin and other CCCH tandem zinc-finger proteins in the regulation of mRNA turnover. *Biochemical Society Transactions* 30: 945-952.

Blackshear PJ, Phillips RS, Lai WS, eds. (2005) Tandem CCCH zinc finger proteins in mRNA binding. Landes Bioscience/Eurekah.com and Kluwer Academic/Plenum Publishers.

Brown RS (2005) Zinc finger proteins: Getting a grip on RNA. *Current Opinion in Structural B*iology 15: 94-98.

Camerini S, Polci ML, Bachi A (2005) Proteomics approaches to study the redox state of cysteine-containing proteins. *Ann. Ist. Super Sanita* 41: 451-457.

Campbell ID, Downing AK (1994) Building protein structure and function from modular units. *Trends in Biotechnology* 12: 168-172.

Carrick DM, Lai WS, Blackshear PJ (2004) The tandem CCCH zinc finger protein tristetraprolin and its relevance to cytokine mRNA turnover and arthritis. *Arthritis Research and Therapy* 6: 248-264.

Chao YH, Kang JM, Sun Y, Yang QC, Wang PQ, Wu MS, Li Y, Long RC, Qin ZH (2009) Molecular cloning and characterization of a novel gene encoding zinc finger protein from Medicago sativa L. *Molecular Biology Reports* 36: 2315-2321.

Cheng Y, Kato N, Wang W, Li J, Chen X (2003) Two RNA binding proteins, HEN4 and HUA1 act in the processing of AGAMOUS pre-mRNA in *Arabidopsis thaliana*. *Developmental Cell* 4: 53-66.

Cho S-H, Lee C-H, Ahn Y, Kim H, Kim H, Ahn C-Y, Yang K-S, Lee S-R (2004) Redox regulation of PTEN and protein tyrosine phosphatases in $H_2O_2$-mediated cell signaling. *FEBS Letters* 560: 7-13.

Delaney KJ, Xu R, Zhang J, Li QQ, Yun K-Y, Falcone DL, Hunt AG (2006) Calmodulin interacts with and regulates the RNA-binding activity of an *Arabidopsis* polyadenylation factor subunit. *Plant Physiol.* 140: 1507-1521.

diTargiani RC, Lee SJ, Wassink S, Michel SL (2006) Functional characterization of iron-substituted tristetraprolin-2D (TTP-2D, NUP475-2D): RNA binding affinity and selectivity. *Biochemistry* 45: 13641-13649.

Dong HK, Yamaguchi S, Lim S, Oh E, Park J, Hanada A, Kamiya Y, Choi G (2008) SOMNUS, a CCCH-type zinc finger protein in *Arabidopsis*, negatively regulates light-dependent seed germination downstream of PIL5. *Plant Cell* 20: 1260-1277.

Gamsjaeger R, Liew CK, Loughlin FE, Crossley M, Mackay JP (2007) Sticky fingers: zinc-fingers as protein-recognition motifs. *Trends in biochemical sciences* 32: 63-70.

Gilbert HF (2006) Molecular and Cellular Aspects of Thiol–Disulfide Exchange. *In* Advances in Enzymology and Related Areas of Molecular Biology. John Wiley and Sons, Inc., pp 69-172.

Grabowska A, Wisniewska A, Tagashira N, Malepszy S, Filipecki M (2009) Characterization of CsSEF1 gene encoding putative CCCH-type zinc finger protein expressed during cucumber somatic embryogenesis. *Journal of Plant Physiology* 166: 310-323.

Guex N, Peitsch MC (1997) SWISS-MODEL and the Swiss-PdbViewer: an environment for comparative protein modeling. *Electrophoresis* 18: 2714-2723.

Guo YH, Yu YP, Wang D, Wu CA, Yang GD, Huang JG, Zheng CC (2009) GhZFP1, a novel CCCH-type zinc finger protein from cotton, enhances salt stress tolerance and fungal disease resistance in transgenic tobacco by interacting with GZIRD21A and GZIPR5. *New Phytologist* 183: 62-75.

Hogg PJ (2003) Disulfide bonds as switches for protein function. *Trends in biochemical sciences* 28: 210-214.

Hudson BP, Martinez-Yamout MA, Dyson HJ, Wright PE (2004) Recognition of the mRNA AU-rich element by the zinc finger domain of TIS11d. *Nature Structural and Molecular Biology* 11: 257-264.

Hunt AG, Xu R, Addepalli B, Rao S, Forbes KP, Meeks LR, Xing D, Mo M, Zhao H, Bandyopadhyay A, Dampanaboina L, Marion A, Von Lanken C, Li QQ (2008) *Arabidopsis* mRNA polyadenylation machinery: Comprehensive analysis of protein-protein interactions and gene expression profiling. *BMC Genomics* 9:220.

Kong Z, Li M, Yang W, Xu W, Xue Y (2006) A novel nuclear-localized CCCH-type zinc finger protein, OsDOS, is involved in delaying leaf senescence in rice. *Plant Physiology* 141: 1376-1388.

Lai WIS, Carballo E, Strum JR, Kennington EA, Phillips RS, Blackshear PJ (1999) Evidence that tristetraprolin binds to AU-rich elements and promotes the deadenylation and destabilization of tumor necrosis factor alpha mRNA. *Molecular and Cellular Biology* 19: 4311-4323.

Lai WS, Kennington EA, Blackshear PJ (2002) Interactions of CCCH zinc finger proteins with mRNA. Non-binding tristetraprolin mutants exert an inhibitory effect on degradation of AU-rich element-containing mRNAs. *Journal of Biological Chemistry* 277: 9606-9613.

Lee SH, Maret W (2001) Redox control of zinc finger proteins: Mechanisms and role in gene regulation. *Antioxidants and Redox Signaling* 3: 531-534.

Lee SJ, Michel SLJ (2010) Cysteine Oxidation Enhanced by Iron in Tristetraprolin, A Zinc Finger Peptide. *Inorg. Chem.* 49: 1211-1219.

Li J, Jia D, Chen X (2001) HUA1, a regulator of stamen and carpel identities in *Arabidopsis*, codes for a nuclear RNA binding protein. *Plant Cell* 13: 2269-2281.

Li ZS, Thomas TL (1998) PEI1, an embryo-specific zinc finger protein gene required for heart-stage embryo formation in *Arabidopsis*. *Plant Cell* 10: 383-398.

Lin PC, Pomeranz MC, Jikumaru Y, Kang SG, Hah C, Fujioka S, Kamiya Y, Jang JC (2011) The *Arabidopsis* tandem zinc finger protein AtTZF1 affects ABA- and GA-mediated growth, stress and gene expression responses. *Plant Journal* 65: 253-268.

Moore M, Ullman C (2003) Recent developments in the engineering of zinc finger proteins. *Briefings in Functional Genomics and Proteomics* 1: 342-355.

Nakashima I, Kato M, Akhand AA, Suzuki H, Takeda K, Hossain K, Kawamoto Y (2002) Redox-linked signal transduction pathways for protein tyrosine kinase activation. *Antioxid Redox Signal* 4: 517-531.

Oh E, Yamaguchi S, Hu J, Yusuke J, Jung B, Paik I, Lee H-S, Sun T-p, Kamiya Y, Choi G (2007) PIL5, a Phytochrome-Interacting bHLH Protein, Regulates Gibberellin Responsiveness by Binding Directly to the GAI and RGA Promoters in *Arabidopsis* Seeds. *The Plant Cell Online* 19: 1192-1208.

Pomeranz MC, Hah C, Lin PC, Kang SG, Finer JJ, Blackshear PJ, Jang JC (2010) The *Arabidopsis* tandem zinc finger protein AtTZF1 traffics between the nucleus and cytoplasmic foci and binds both DNA and RNA. *Plant Physiology* 152: 151-165.

Rao S, Dinkins RD, Hunt AG (2009) Distinctive interactions of the *Arabidopsis* homolog of the 30 kD subunit of the cleavage and polyadenylation specificity factor (AtCPSF30) with other polyadenylation factor subunits. *BMC Cell Biol.* 10: 51.

Reddy ASN, Ali GS, Celesnik H, Day IS (2011) Coping with Stresses: Roles of Calcium- and Calcium/Calmodulin- Regulated Gene Expression. *The Plant Cell* 23:2010-2032.

Schmitz RJ, Hong L, Michaels S, Amasino RM (2005) FRIGIDA-ESSENTIAL 1 interacts genetically with FRIGIDA and FRIGIDA-LIKE 1 to promote the winter-annual habit of *Arabidopsis thaliana*. *Development* 132: 5471-5478.

Schwede T, Kopp Jr, Guex N, Peitsch MC (2003) SWISS-MODEL: an automated protein homology-modeling server. *Nucleic Acids Research* 31: 3381-3385.

Sun J, Jiang H, Xu Y, Li H, Wu X, Xie Q, Li C (2007) The CCCH-type zinc finger proteins AtSZF1 and AtSZF2 regulate salt stress responses in *Arabidopsis*. *Plant and Cell Physiology* 48: 1148-1158.

Takatsuji H (1998) Zinc-finger transcription factors in plants. *Cellular and Molecular Life Sciences* 54: 582-596.

Wang D, Guo Y, Wu C, Yang G, Li Y, Zheng C (2008) Genome-wide analysis of CCCH zinc finger family in *Arabidopsis* and rice. BMC Genomics 9:44.

Wang L, Xu Y, Zhang C, Ma Q, Joo SH, Kim SK, Xu Z, Chong K (2008) OsLIC, a novel CCCH-type zinc finger protein with transcription activation, mediates rice architecture via brassinosteroids signaling. *PLoS ONE* 3:e3521.

Yamamuro C, Ihara Y, Wu X, Noguchi T, Fujioka S, Takatsuto S, Ashikari M, Kitano H, Matsuoka M (2000) Loss of Function of a Rice brassinosteroid insensitivel Homolog Prevents Internode Elongation and Bending of the Lamina Joint. *The Plant Cell Online* 12: 1591-1606.

Yang T, Poovaiah BW (2002) Hydrogen peroxide homeostasis: Activation of plant catalase by calcium/calmodulin. *Proceedings of the National Academy of Sciences* 99: 4097-4102.

Zhang J, Addepalli B, Yun K-Y, Hunt AG, Xu R, Rao S, Li QQ, Falcone DL (2008) A polyadenylation factor subunit implicated in regulating oxidative signaling in *Arabidopsis thaliana*. *PLoS ONE* 3:e2410.

In In: Zinc Fingers
Editors: R. Ciofani and L. Makrlik
ISBN 978-1-62100-230-7


*Chapter 6*

# New Functions of Zinc Fingers Revealed by Substitution of Conserved Residues

***Miki Imanishi*[1] *and Yukio Sugiura*[2]**
[1]Institute for Chemical Research, Kyoto University, Uji, Kyoto 611-0011, Japan
[2]Fuculty of Pharmaceutical Sciences, Doshisha Women's University, Kyotanabe 610-0395, Japan

## Abstract

The C2H2-type zinc finger motif is one of the most major DNA binding motifs in human transcription factors. The motif commonly contains the sequence of the type, (Tyr, Phe)-X-Cys-$X_{2-4}$-Cys-$X_3$-Phe-$X_5$-Leu-$X_2$-His-$X_{3-5}$-His, where X represents relatively non-conserved amino acids. Each finger folds into a compact $\beta\beta\alpha$ structure by coordination of a zinc ion with the conserved two cysteine (C) and two histidine(H) residues in a tetrahedral fashion. This fold is stabilized by hydrophobic interactions formed by the conserved hydrophobic amino acid residues. Though the Zn(II) ligands, two cysteine and two histidine residues, and the hydrophobic amino acid residues are highly conserved among the C2H2-type zinc fingers, their contribution to structure and functions is not clearly understood. We have created zinc finger mutants, in which the highly conserved amino acid residues are substituted to other amino acid residues. Mutation of the conserved hydrophobic amino acids on the DNA recognition helix or β-turn clarified the relationship between formation of a hydrophobic core and DNA binding function. On the other hand, ligand substituted HHHH-, CHHH-, and HCHH-type zinc fingers, in which cysteine residue(s) is mutated to histidine residue(s), showed hydrolytic ability and cleaved DNA in a sequence specific manner. They would be a new type of nucleases. In addition, we have successfully created the Zn(II)-dependent DNA binding domain, CDHH zinc finger, by reducing the Zn(II) binding affinity via ligand substitution of the second cysteine by aspartic acid. An artificial transcription factor with a CDHH-type zinc finger as a DNA binding domain showed Zn(II)-responsive transcriptional activity inside living cells. Here, we review the new functions of zinc fingers revealed by substitution of highly conserved residues.

# 1. Conserved Amino Acid Residues in C2H2-Zinc Finger Motifs

The C2H2 zinc finger motif is one of the most common DNA binding motifs in eukaryotes. The motifs were first discovered in the Xenopus 5S RNA gene-specific transcription factor, TFIIIA, as repeated motifs of 30 amino acids (Miller et al., 1985). The C2H2 zinc finger motifs have a highly conserved sequence, (Tyr, Phe)-X-Cys-$X_{2\text{-}4}$-Cys-$X_3$-Phe-$X_5$-Leu-$X_2$-His-$X_{3\text{-}5}$-His, where X represents relatively non-conserved amino acids (Figure 1). A Zn(II) ion binds to two cysteine and two histidine residues(Berg, 1988), which are essential to form the ββα secondary structure of a zinc finger domain (Figure 2)(Lee et al., 1989; Neuhaus et al., 1992). The conserved cysteine residues are within the β-turn and the histidine residues are within the α-helix. The folding of a zinc finger motif is also stabilized by hydrophobic interactions formed by the conserved hydrophobic amino acid residues (Thy and Phe in the β-turn and Leu in the α-helix). Berg and coworkers synthesized a peptide, Lys-Tyr-Ala-Cys-$Ala_2$-Cys-$Ala_3$-Phe-Ala-Ala-Lys-Ala-Ala-Leu-$Ala_2$-His-$Ala_3$-His-Ala-Lys, and showed that the metal-binding and conserved hydrophobic residues are necessary and sufficient to determine the structure of a C2H2 zinc finger peptide (Michael et al., 1992). The importance of the hydrophobic core in the structural stability was proved by *de novo* design of the zinc finger peptides that assumed a ββα fold without a Zn(II) ion but with a minimal hydrophobic core by juxtaposing hydrophobic amino acid residues.

Imperiali and coworkers designed a 23-residue peptide folding into the ββα fold without a Zn(II) ion (Struthers et al., 1996).

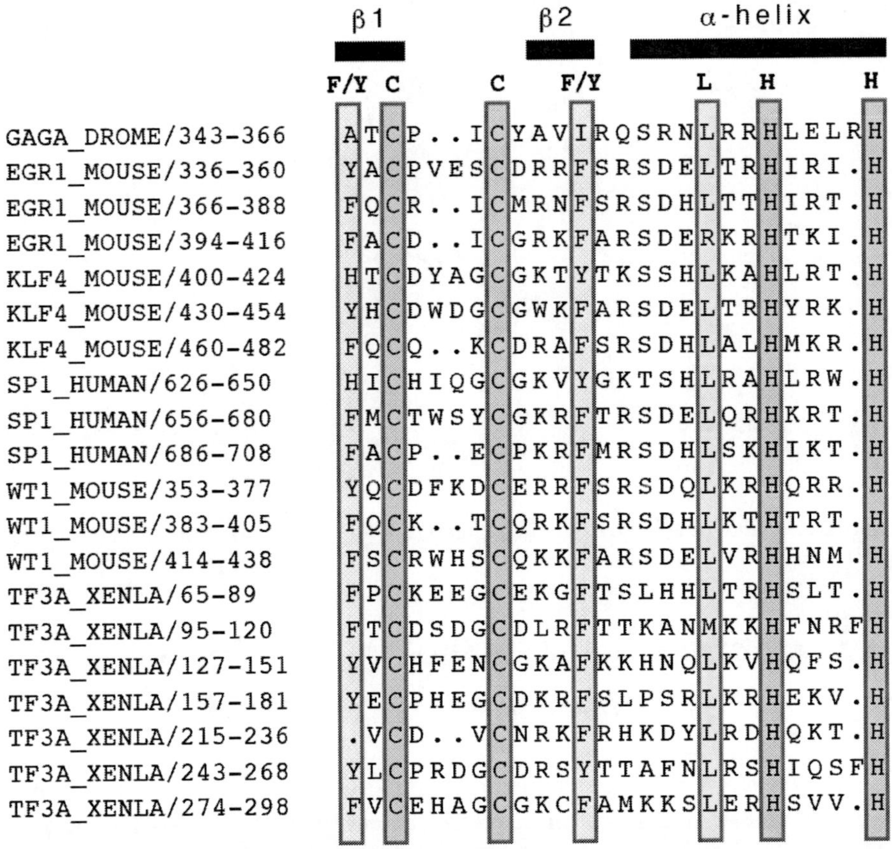


Figure 1. Sequence alignment of the representative CCHH zinc finger domains. The highly conserved amino acid residues are highlighted. As shown in this alignment, the consensus amino acid sequence of the CCHH zinc finger domains is (Tyr, Phe)-X-Cys-$X_{2\text{-}4}$-Cys-$X_3$-Phe-$X_5$-Leu-$X_2$-His-$X_{3\text{-}5}$-His. The positions of the DNA recognition helix and two β-strands are indicated by the black bars above the sequences.

Similarly, a 28-residue peptide with a defined ββα fold without a Zn(II) ion was designed using a computational design algorithm by Mayo and coworkers (Dahiyat and Mayo, 1997; Dahiyat et al., 1997). However, the DNA binding ability of these designed zinc finger peptides was not examined. There have been a few reports that investigated the relationships between the stability of the hydrophobic core and DNA binding function (Dhanasekaran et al., 2007; Dhanasekaran et al., 2008).

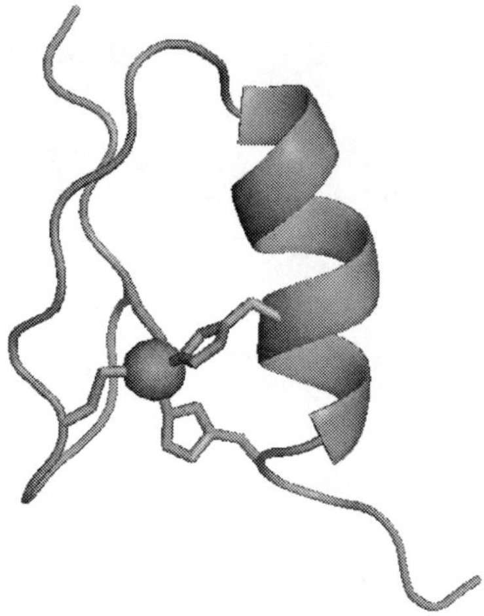

Figure 2. Structure of a CCHH zinc finger motif, Sp1 finger 2 (PDB; 1SP2, Narayan et al., 1997).

## 2. GAGA Transcription Factor Contains a Single Zinc Finger Motif

Usually DNA binding domains containing C2H2 zinc finger motifs have varying numbers of arrays of these motifs, at least two zinc fingers for specific DNA recognition, *i.e.*, Sp1 and Zif268 contain three zinc finger motifs, TFIIIA nine fingers, GLI five fingers, et al.(Elrod-Erickson et al., 1998; Elrod-Erickson and Pabo, 1999; Elrod-Erickson et al., 1996; Kuwahara and Coleman, 1990; Narayan et al., 1997; Oka et al., 2004; Pavletich and Pabo, 1991; Pavletich and Pabo, 1993; Wuttke et al., 1997). On the other hand, the GAGA factor of *Drosophila melanogaster* uses a single C2H2 zinc finger for specific DNA binding (Pedone et al., 1996). The minimal DNA binding domain of the GAGA transcription factor (GAGA-DBD) has 63 residues in length and binds to the DNA sequence containing GAGAGAG. The GAGA zinc finger core is a 26-residue peptide and the N-terminal basic region containing BR1 and BR2 is necessary for specific DNA binding in addition to the zinc finger core (Figure 3).

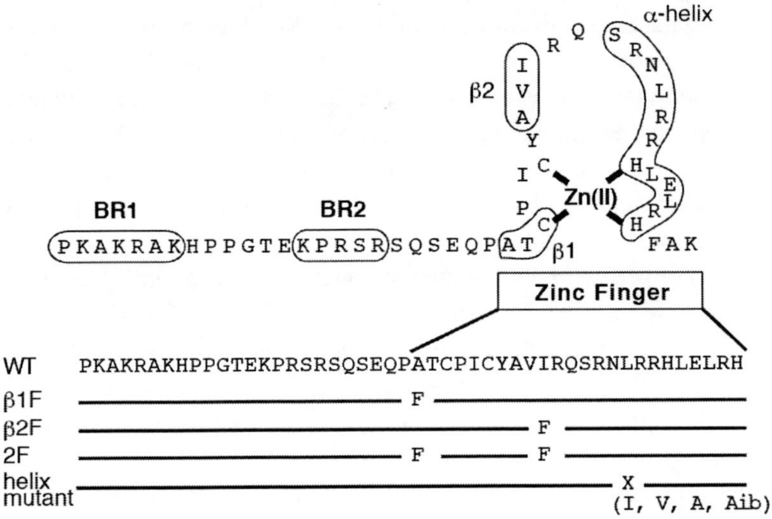


Figure 3. Amino acid sequences of the wild type and mutant GAGA zinc finger peptides. The N-terminal tail containing BR1 and BR2 basic regions is important for DNA binding. In GAGA zinc finger, the conserved aromatic amino acid residues at the β1 and β2 strands are Ala and Ile, respectively. These residues were substituted by conserved phenylalanine. On the contrary, the conserved leucine residue on the α-helix was substituted by Ile, Val, Ala, and Aib.

The NMR structure of the GAGA-DBD/DNA complex revealed that the structure as well as the DNA binding mode is very similar to other classical C2H2 zinc fingers (Omichinski et al., 1997). From these properties, the GAGA-DBD is an attractive domain to directly investigate correlations between the zinc finger conformation and DNA binding function.

## 3. Effects of Hydrophobic Amino Acid Residues in the β-Hairpin of the GAGA Zinc Finger

The amino acid preference in the β-hairpin region of the GAGA zinc finger is different from other C2H2 zinc finger proteins. The first position of β-strand1 (β1) and the third position of β-strand2 (β2) are occupied by alanine and isoleucine, respectively, in the GAGA zinc finger, though phenylalanine, thyrosine, and histidine are usually preferred at the first position of β1, and phenylalanine is highly conserved at the third position of β2. It was shown that the thermodynamic or dynamic stability of the folded form of individual typical zinc fingers is significantly decreased by the substitution of a consensus phenylalanine residue with non-aromatic amino acid residues in the β-hairpin despite retaining the ββα fold (Lachenmann et al., 2002; Mortishire-Smith et al., 1992; Weiss and Keutmann, 1990). On the other hand, the non-aromatic amino acid residues of the GAGA zinc finger were substituted by aromatic amino acid residues, and then their structures and DNA binding abilities were investigated by Dhanasekaranet al. (Dhanasekaran et al., 2007). The alanine residue of β1 and the isoleucine residue of β2 were substituted to the conserved phenylalanine residue to create the β1F and β2F peptides, respectively. Both of them were substituted yielding the 2F peptide (Figure 3).

The UV-vis absorption spectra in the presence of Co(II) ion, a spectroscopic probe for the zinc site, showed that the mutation in the β-hairpin region does not alter the coordination geometry of the metal binding. In addition, all the mutant peptides display a zinc finger-specific CD spectrum in the presence of Zn(II) ion. The thermal denaturing experiment also suggests that the mutation does not affect the thermal stability of the GAGA peptide. Namely, the mutation of aliphatic amino acid residue(s) to conserved aromatic amino acid(s), phenylalanine, does not clearly change the Zn(II) binding and folding properties of the GAGA zinc finger peptide. On the other hand, the gel mobility shift assay and isothermal calorimetry (ITC) showed that none of the mutants are able to bind the cognate DNA substrate, although the mutation is confined only to the β-hairpin region, not directly involving in DNA recognition. The loss in the DNA binding activity could be explained by the change in the geometric requirements of the DNA recognition due to the variation in the side-chain packing or in the minimal hydrophobic core (Weiss, M.A., and Keutmann, H. 1990, Mortishire-Smith, R.J., et al.., 1992, Kochoyan, M., 1991).

## 4. Effects of the Conserved Leucine Residue in the α-Helix of the GAGA Zinc Finger

At the fourth position of the recognition helix of a typical C2H2 zinc finger motif, a leucine residue is highly conserved. In order to investigate the contribution of the leucine residue on the stability of the hydrophobic core and

DNA binding activity, a set of four GAGA zinc finger mutant peptides were prepared by substituting the leucine residue with other aliphatic residues, *i.e.*, isoleucine (Ile), valine (Val), 2-aminoisobutyric acid (Aib), and alanine (Ala), having different degrees of bulkiness and hydrophobicities of the side chains (Figure 3) (Dhanasekaran et al., 2008).

The UV-vis absorption spectra in the presence of Co(II) ion showed that the substitution of the conserved leucine residue in the α-helix region does not alter the coordination geometry of the metal binding. CD spectra indicate that all peptides except for the Ala mutant form a similar conformation to wild type (Leu). The thermal stability judged from CD spectral changes was also comparable independent of the length of the side chains. These results indicate that no gross changes in the secondary structures of the GAGA mutant peptides are induced by the substitution of the conserved leucine residue with other aliphatic amino acid residues. On the other hand, the DNA binding affinity decreased as the side chain length of the amino acid decreased. The Aib mutant did not bind to DNA, in spite of the similarity in the secondary structure to other functional peptides. The reason was estimated by variable temperature NMR experiments, showing slight changes in the amide proton signals for the wild type and Val peptides but temperature-dependent significant chemical shifts for the Aib mutant. The peptide modeling also suggests that the loss of DNA binding activity of the Aib mutant is possibly due to the decreased bulkiness of the side chain not being sufficient for formation of a stable hydrophobic core. The substitution of the conserved leucine residue in the recognition helix of C2H2 zinc finger motif demonstrates that the bulkiness of the leucine residue is important for the formation of the well-packed hydrophobic core and proper ternary structure, and that stability of the hydrophobic core influences the DNA binding function of the zinc finger domain.

# 5. Contribution of Zinc Finger Ligands

In order to understand the contribution of individual zinc-ligating amino acid residues for coupling between zinc binding and protein folding in zinc finger domains, zinc finger mutant peptides were created (Cook et al., 1994; Hori et al., 2000; Merkle and Berg, 1991; Nomura and Sugiura, 2002; Simpson et al., 2003). From these reports, it was demonstrated that each ligand does not contribute equally to helix-formation and coordination geometry in the zinc finger peptide. Especially, the third zinc ligand (*i.e.*, first histidine) was indicated to be essential to form a compact ββα zinc finger structure (Nomura and Sugiura, 2002). On the other hand, the sequence databases examined by Simpson et al. revealed that there are a number of proteins that contain sequences corresponding to one or more classical zinc finger motifs, with the exception that the final zinc-lignad in the last zinc finger is neither cysteine nor histidine(Simpson et al., 2003). In fact, CCHX peptides, in which the last zinc ligand was substituted, showed zinc-responsive helix formation and DNA binding ability (Cook et al., 1994; Simpson et al., 2003). Truncation of the second histidine residue also presented a metal binding ability, namely the absorption spectrum of the complex formed between this peptide and Co(II) showed tetrahedral coordination of a metal ion, though the metal binding affinity was weaker than wild type (Merkle and Berg, 1991). Although care should be taken before presuming the functionality of CCHX zinc finger proteins, these reports suggest that the final zinc ligand (the second histidine) is not an essential factor of classical zinc finger proteins.

# 6. The Hydrolysis Activity of Ligand Substituted Zinc Finger Peptides

Generally zinc can play a structural or catalytic role in metalloproteins(Auld, 2001). As described before, Zn(II) plays a structural role in C2H2 zinc finger proteins, *i.e.*, Zn(II) is necessary for the compact globular ββα zinc finger structure. On the other hand, a catalytic zinc ion directly interacts with substrate molecules.

One of the differences between structural and catalytic zinc ions is the zinc coordination geometry. A catalytic zinc site contains a vacant site(s) and the zinc-bound water is a critical component for a catalytic zinc site (Vallee and Auld, 1992a; Vallee and Auld, 1992b), while a zinc ion tetrahedrally coordinates and there is no vacant site in a classical C2H2 zinc finger motif. Therefore, in order to generate a vacant site(s), a zinc ligand(s) was substituted with various amino acids (Negi et al., 2004; Nomura and Sugiura, 2004a). The hydrolysis ability of the zinc finger mutants was examined by using 4-nitrophenyl acetate (NA) as the substrate.

Among the examined mutants, zinc finger peptides that have more than three histidine residues as zinc ligands, the CHHH-, HCHH-, and HHHH-type zinc finger peptides, showed high hydrolysis ability. These mutant peptides still showed the DNA binding activity to the target DNA sequence. Additionally, the mutant zinc finger domain of transcription factor Sp1, in which all of the three C2H2-type zinc finger motifs were exchanged to the HHHH-type zinc finger motifs, was created and investigated the sequence specific DNA hydrolysis activity (Nomura and Sugiura, 2004b).

It was shown that the HHHH-Sp1 protein bound to the target GC-box DNA and cleaved the plasmid DNA containing the GC-box sequence. So far, a lot of zinc finger nucleases have been developed by fusing zinc finger DNA binding domains with the *Fok*I nuclease domain(Jamieson et al., 2003; Klug, 2010; Urnov et al., 2005). The ligand substituted HHHH-type zinc finger peptides would be a new type nuclease.

# 7. Zn(II)-Responsive DNA Binding Activity of an Artificial Ligand-Substituted Zinc Finger Peptide

Zn(II) binding to a C2H2 zinc finger peptide is necessary for the proper folding of the zinc finger peptide to a ββα structure and for DNA binding activity. Generally the binding between a typical C2H2 zinc finger peptide and Zn(II) is very stable (Reddi et al., 2007; Seneque and Latour, 2010). So it is usually difficult to control the DNA binding activity by changing the concentration of Zn(II) because a typical C2H2 zinc finger peptide always holds Zn(II) and presents DNA binding ability under a physiological condition. In order to create a transcriptional switch system controlled by Zn(II) ions, a CDHH zinc finger peptide, in which the second ligand cysteine is substituted with an aspartic acid residue, was designed (Imanishi et al., 2010). The CDHH peptide showed about 500-fold lower Zn(II) binding affinity than the wild C2H2-type. In order to examine the DNA binding activity of the ligand substituted peptide, the second zinc finger motif of the mouse transcription factor Zif268 DNA binding domain was replaced to the CDHH-type zinc finger motif (Figure 4). In the presence of Zn(II) ions, the CDHH peptide showed strong DNA binding ability to the cognate site. In the presence of EDTA, a Zn(II) chelator, the CDHH peptide did not bind to DNA, while the wild C2H2 peptide showed strong DNA binding activity.

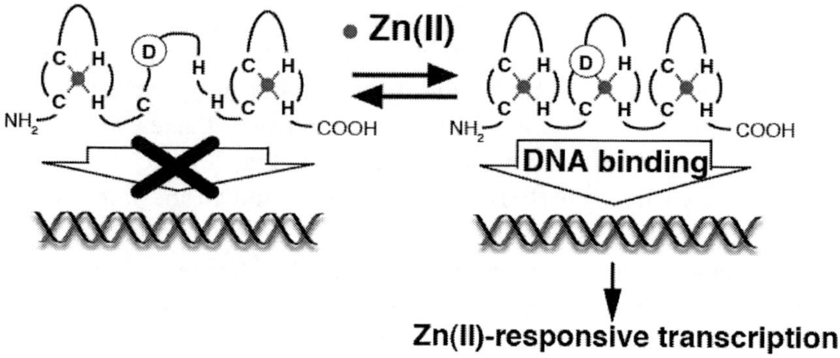


Figure 4. Schematic representation of the Zn(II)-responsive DNA binding and transcriptional control of an artificial zinc finger protein containing a CDHH-type zinc finger motif. Under a low Zn(II) concentration, the CDHH zinc finger will not bind to a zinc ion yielding unfolded structure (left). In the presence of Zn(II), the CDHH zinc finger binds to a zinc ion and presents DNA binding activity.

The regulation of the DNA binding activity by Zn(II) was also demonstrated within living cells. An artificial transcription factor containing a CDHH-type DNA binding domain, SV40 large T antigen derived nuclear localization signal (NLS), and a transcriptional activation domain (VP64) was designed and expressed within cells. The artificial transcription factor activated the reporter vector, in which four tandemly repeated zinc finger binding sequences are inserted at the promoter region of a luciferase gene, in response to the concentration of Zn(II) ion added to the medium. So far, design of zinc finger proteins with new DNA binding specificity has been focused on the recognition helix of zinc finger motifs (Maeder et al., 2009; Rebar and Pabo, 1994; Segal et al., 1999). The results of the CDHH zinc finger indicate that the each zinc ligand could be a design target for artificial DNA binding zinc finger proteins.

## Summary

The design of artificial zinc finger proteins has shown tremendous developments in this decade. Though highly conserved amino acid residues in the C2H2 zinc finger motifs have not changed so much, the results described in this review show the importance of the zinc finger backbone structures as well as the DNA recognition helix. In addition, the fact that new functions, *i.e.* hydrolysis activity and Zn(II)-responsive DNA binding activity, can be obtained by ligand substitutions, will open the new design strategy of artificial functional zinc finger proteins.

# References


Auld DS (2001) Zinc coordination sphere in biochemical zinc sites. *Biometals.* 14(3-4): 271-313.

Berg JM (1988) Proposed structure for the zinc-binding domains from transcription factor IIIA and related proteins. *Proc. Natl. Acad. Sci. U S A* 85(1): 99-102.

Cook WJ, Mosley SP, Audino DC, Mullaney DL, Rovelli A, Stewart G, Denis CL (1994) Mutations in the zinc-finger region of the yeast regulatory protein ADR1 affect both DNA binding and transcriptional activation. *J. Biol. Chem.* 269(12): 9374-9379.

Dahiyat BI, Mayo SL (1997) De novo protein design: fully automated sequence selection. *Science* 278(5335): 82-87.

Dahiyat BI, Sarisky CA, Mayo SL (1997) De novo protein design: towards fully automated sequence selection. *J. Mol. Biol.* 273(4): 789-796.

Dhanasekaran M, Negi S, Imanishi M, Sugiura Y (2007) DNA-Binding ability of GAGA zinc finger depends on the nature of amino acids present in the beta-hairpin. *Biochemistry* 46(25): 7506-7513.

Dhanasekaran M, Negi S, Imanishi M, Suzuki M, Sugiura Y (2008) Effects of bulkiness and hydrophobicity of an aliphatic amino acid in the recognition helix of the GAGA zinc finger on the stability of the hydrophobic core and DNA binding affinity. *Biochemistry* 47(45): 11717-11724.

Elrod-Erickson M, Benson TE, Pabo CO (1998) High-resolution structures of variant Zif268-DNA complexes: implications for understanding zinc finger-DNA recognition. *Structure* 6(4): 451-464.

Elrod-Erickson M, Pabo CO (1999) Binding studies with mutants of Zif268. Contribution of individual side chains to binding affinity and specificity in the Zif268 zinc finger-DNA complex. *J. Biol. Chem.* 274(27): 19281-19285.

Elrod-Erickson M, Rould MA, Nekludova L, Pabo CO (1996) Zif268 protein-DNA complex refined at 1.6 A: a model system for understanding zinc finger-DNA interactions. *Structure* 4(10): 1171-1180.

Hori Y, Suzuki K, Okuno Y, Nagaoka M, Futaki S, Sugiura Y (2000) Artificial zinc finger peptide containing a novel His4 domain. *J Am. Chem. Soc.* 122(32): 7648-7653.

Imanishi M, Nakaya T, Morisaki T, Noshiro D, Futaki S, Sugiura Y (2010) Metal-stimulated regulation of transcription by an artificial zinc-finger protein. *Chembiochem.* 11(12): 1653-1655.

Jamieson AC, Miller JC, Pabo CO (2003) Drug discovery with engineered zinc-finger proteins. *Nat. Rev. Drug Discov.* 2(5): 361-368.

Klug A (2010) The discovery of zinc fingers and their applications in gene regulation and genome manipulation. *Annu. Rev. Biochem.* 79: 213-231.

Kuwahara J, Coleman JE (1990) Role of the zinc(II) ions in the structure of the three-finger DNA binding domain of the Sp1 transcription factor. *Biochemistry* 29(37): 8627-8631.

Lachenmann MJ, Ladbury JE, Phillips NB, Narayana N, Qian X, Weiss MA (2002) The hidden thermodynamics of a zinc finger. *J. Mol. Biol.* 316(4): 969-989.

Lee MS, Gippert GP, Soman KV, Case DA, Wright PE (1989) Three-dimensional solution structure of a single zinc finger DNA-binding domain. *Science* 245(4918): 635-637.

Maeder ML, Thibodeau-Beganny S, Sander JD, Voytas DF, Joung JK (2009) Oligomerized pool engineering (OPEN): an 'open-source' protocol for making customized zinc-finger arrays. *Nat. Protoc.* 4(10): 1471-1501.

Merkle DL, Berg JM (1991) Metal requirements for nucleic acid binding proteins. *Methods Enzymol.* 208: 46-54.

Michael SF, Kilfoil VJ, Schmidt MH, Amann BT, Berg JM (1992) Metal binding and folding properties of a minimalist Cys2His2 zinc finger peptide. *Proc. Natl. Acad. Sci. U S A* 89(11): 4796-4800.

Miller J, McLachlan AD, Klug A (1985) Repetitive zinc-binding domains in the protein transcription factor IIIA from Xenopus oocytes. *EMBO J.* 4(6): 1609-1614.

Mortishire-Smith RJ, Lee MS, Bolinger L, Wright PE (1992) Structural determinants of Cys2His2 zinc fingers. *FEBS Lett.* 296(1): 11-15.

Narayan VA, Kriwacki RW, Caradonna JP (1997) Structures of zinc finger domains from transcription factor Sp1. Insights into sequence-specific protein-DNA recognition. *J. Biol. Chem.* 272(12): 7801-7809.

Negi S, Itazu M, Imanishi M, Nomura A, Sugiura Y (2004) Creation and characteristics of unnatural CysHis(3)-type zinc finger protein. *Biochem. Biophys. Res. Commun.* 325(2): 421-425.

Neuhaus D, Nakaseko Y, Schwabe JW, Klug A (1992) Solution structures of two zinc-finger domains from SWI5 obtained using two-dimensional 1H nuclear magnetic resonance spectroscopy. A zinc-finger structure with a third strand of beta-sheet. *J. Mol. Biol.* 228(2): 637-651.

Nomura A, Sugiura Y (2002) Contribution of individual zinc ligands to metal binding and peptide folding of zinc finger peptides. *Inorg. Chem.* 41(14): 3693-3698.

Nomura A, Sugiura Y (2004a) Hydrolytic reaction by zinc finger mutant peptides: successful redesign of structural zinc sites into catalytic zinc sites. *Inorg. Chem.* 43(5): 1708-1713.

Nomura A, Sugiura Y (2004b) Sequence-selective and hydrolytic cleavage of DNA by zinc finger mutants. *J. Am. Chem. Soc.* 126(47): 15374-15375.

Oka S, Shiraishi Y, Yoshida T, Ohkubo T, Sugiura Y, Kobayashi Y (2004) NMR structure of transcription factor Sp1 DNA binding domain. *Biochemistry* 43(51): 16027-16035.

Omichinski JG, Pedone PV, Felsenfeld G, Gronenborn AM, Clore GM (1997) The solution structure of a specific GAGA factor-DNA complex reveals a modular binding mode. *Nat. Struct. Biol.* 4(2): 122-132.

Pavletich NP, Pabo CO (1991) Zinc finger-DNA recognition: crystal structure of a Zif268-DNA complex at 2.1 A. *Science* 252(5007): 809-817.

Pavletich NP, Pabo CO (1993) Crystal structure of a five-finger GLI-DNA complex: new perspectives on zinc fingers. *Science* 261(5129): 1701-1707.

Pedone PV, Ghirlando R, Clore GM, Gronenborn AM, Felsenfeld G, Omichinski JG (1996) The single Cys2-His2 zinc finger domain of the GAGA protein flanked by basic residues is sufficient for high-affinity specific DNA binding. *Proc. Natl. Acad. Sci. U S A* 93(7): 2822-2826.

Rebar EJ, Pabo CO (1994) Zinc finger phage: affinity selection of fingers with new DNA-binding specificities. *Science* 263(5147): 671-673.

Reddi AR, Guzman TR, Breece RM, Tierney DL, Gibney BR (2007) Deducing the energetic cost of protein folding in zinc finger proteins using designed metallopeptides. *J. Am. Chem. Soc.* 129(42): 12815-12827.

Segal DJ, Dreier B, Beerli RR, Barbas CF, 3rd (1999) Toward controlling gene expression at will: selection and design of zinc finger domains recognizing each of the 5'-GNN-3' DNA target sequences. *Proc. Natl. Acad. Sci. U S A* 96(6): 2758-2763.

Seneque O, Latour JM (2010) Coordination properties of zinc finger peptides revisited: ligand competition studies reveal higher affinities for zinc and cobalt. *J. Am. Chem. Soc.* 132(50): 17760-17774.

Simpson RJ, Cram ED, Czolij R, Matthews JM, Crossley M, Mackay JP (2003) CCHX zinc finger derivatives retain the ability to bind Zn(II) and mediate protein-DNA interactions. *J. Biol. Chem.* 278(30): 28011-28018.

Struthers MD, Cheng RP, Imperiali B (1996) Design of a monomeric 23-residue polypeptide with defined tertiary structure. *Science* 271(5247): 342-345.

Urnov FD, Miller JC, Lee YL, Beausejour CM, Rock JM, Augustus S, Jamieson AC, Porteus MH, Gregory PD, Holmes MC (2005) Highly efficient endogenous human gene correction using designed zinc-finger nucleases. *Nature* 435(7042): 646-651.

Vallee BL, Auld DS (1992a) Active zinc binding sites of zinc metalloenzymes. *Matrix Suppl.* 1: 5-19.

Vallee BL, Auld DS (1992b) Functional zinc-binding motifs in enzymes and DNA-binding proteins. *Faraday Discuss*(93): 47-65.

Weiss MA, Keutmann HT (1990) Alternating zinc finger motifs in the male-associated protein ZFY: defining architectural rules by mutagenesis and design of an "aromatic swap" second-site revertant. *Biochemistry* 29(42): 9808-9813.

Wuttke DS, Foster MP, Case DA, Gottesfeld JM, Wright PE (1997) Solution structure of the first three zinc fingers of TFIIIA bound to the cognate DNA sequence: determinants of affinity and sequence specificity. *J. Mol. Biol.* 273(1): 183-206.

In In: Zinc Fingers
Editors: R. Ciofani and L. Makrlik
ISBN 978-1-62100-230-7


*Chapter 7*

# Role of Transcription Factor Sp1 in Expression of Kidney Glucose Transporters in Health and Disease

***Niloofar M. Tabatabai*[*1,2], *Rajendra K. Kothinti*[1,2] *and Amy B. Blodgett*[2]**

[1]Division of Endocrinology, Metabolism and Clinical Nutrition and
[2]Kidney Disease Center, Medical College of Wisconsin,
Milwaukee, Wisconsin, USA

## Abstract

Sp1 is a member of the zinc finger (ZF) family of transcription factor proteins which bind to GC-rich sequences, known as GC-boxes. Sp1 is a ubiquitous transcription factor and plays role in the regulation of expression of many genes. It has been shown *in vitro* that the promoter activity of human *SGLT1* gene is regulated by Sp1. This transcription factor also binds to a GC-box upstream of mouse *SGLT2* promoter. These

* Corresponding author: Niloofar M. Tabatabai, Ph.D., Kidney Disease Center. Medical College of Wisconsin, 8701 Watertown Plank Road, Milwaukee, Wisconsin 53226, Phone: 414-456-8452, Email: ntabatab@mcw.edu

genes encode proteins that are involved in transport of extracellular glucose into the cells. In the kidney, SGLT1 and SGLT2 are expressed in proximal tubule cells of the nephron and play crucial roles in reabsorption of glucose from glomerular filtrate. Chronic exposure of human to cadmium results in accumulation of this toxic metal in the kidney cortex which can cause down-regulation of glucose reabsorption and development of the symptom of glucosuria. The mechanism of cadmium-induced glucosuria may involves the inhibitory effect of this metal on the DNA binding activity of Sp1 that can mediate the transcriptional down-regulation of *SGLT1* and *SGLT2* gene expression.

# Transcription Factor Sp1

Human transcription factor Sp1 is a 785 amino acid long protein and its sequence is highly conserved among different species (Figure 1). Sp1 belongs to the family of zinc finger (ZF) proteins which are characterized by their conserved C-terminus three tandem $Cys_2His_2$ ($C_2H_2$) domains [17]. The two Cys and the two His residues in each ZF domain of Sp1 coordinate one zinc (Zn) ion which is required for Sp1 DNA binding activity [22,35]. Sp1 preferentially binds to the GC-rich regions, known as GC-box, which have the consensus sequence of GGGGCGGGG (31). Sp1 can also bind to the GT-rich regions with the consensus sequence of GGTGTGGGG. The NMR structural studies have shown that ZF domains of Sp1 participate in binding of this transcription factor to the GC-box sequences (Figure 2) [26,29].

Sp1 is ubiquitously expressed in mammalian cells [23] and plays role as an activator of gene expression [5]. This transcription factor is involved in the regulation of expression of many genes [45] and its crucial role has been demonstrated by showing that mice embryos homozygous for deletion of *Sp1* gene are severely retarded in growth and do not survive the gestation period [5]. Sp1 can undergo different post-translational modifications that include phosphorylation [8], glycosylation [14,5], SUMOylation (44), and acetylation [53]; such modifications can alter the activity of Sp1.

# Sp1 in Kidney Cells

Expression of Sp1 has been shown in the kidneys of mouse [12] and rat [57]. HK-2 is a proximal tubule cell line that is derived from a human normal kidney cortex [37], and HEK-293 is a human embryonic kidney cell line (11). Sp1 is expressed in HK-2 cells [25,57] and its activity has been shown in HEK-293 cell line [27].

```
                   .         .         .         .         .
Human     MSDQDHSMDEMTAVVKIEKGVGGNNGGNGNGG-GAFSQARSSSTGSSSST-GGGG-QESQ 57
Chimp     MSDQDHSMDEMTAVVKIEKGVGGNNGGNGNGG-GAFSQARSSSTGSSSST-GGGGGQESQ 58
Cow       MSDQDHSMDEMTAVVKIEKGVGGNNGGNGNGS-GAFSQARSSSAGSSSSS--GGGGQESQ 57
Dog       -------MDEMTAVVKIEKGVGGNNGGNGNGG-GAFSQARSSSTGSSSS---GGGGQESQ 49
Pig       MSDQDHSMDEMTAVVKIEKGVGGNNGGSGNG--GAFSQARSSSTGSSSSGG-GGGGQESQ 57
Rabbit    MSDQDHSMDEMTAVVKIEKGVGGNNGGNGNGG-GAFSQARSSSTGSSSSSGGGGGGQESQ 59
Mouse     MSDQDHSMDEVTAVVKIEKDVGGNNGGSGNGGAAFSQTRSSSTGSSSSS-GGGGGQESQ 59
Rat       MSDQDHSMDEVTA-VKIEKGVGGNNGGSGNGGAAFSQTRSSSTGSSSSS-GGGGGQESQ 58
                 ***:** *****.*******.***  .****:****:*****   *** ****

                   .         .         .         .         .
Human     PSPLALLAATCSRIESPNENSNNSQGPSQSGGTGELDLTATQLSQGANGWQIISSSSGAT 117
Chimp     PSPLALLAATCSRIESPNENSNNSQGPSQSGGTGELDLTATQLSQGANGWQIISSSSGAT 118
Cow       PSPLALLAATCSRIESPNENSNNSQGPSQSGGTGELDLTATQLSQGANGWQIISSSSGAT 117
Dog       PSPLALLAATCSRIESPNENSNNSQGPSQSGGTGELDLTATQLSQGANGWQIISSSSGAT 109
Pig       PSPLALLAATCSRIESPNENSNNSQGPGQSGSTGELDLTATQLSQSANGWQIISSSSGAT 117
Rabbit    PSPLALLAATCSRIESPNENSNNSQGPSQSGGTGELDLTAAQLSQGANGWQIISSSSGAT 119
Mouse     PSPLALLAATCSRIESPNENSNNSQGPSQSGGTGELDLTAAQLSQGANGWQIISSSSGAT 119
Rat       PSPLALLAATCSRIESPNENSNNSQGPSQSGGTGELDLTATQLSQGANGWQIISSSSGAT 118
          ***************************.***.********:****.**************

                   .         .         .         .         .
Human     PTSKEQSGSSTNGSNGSESSKNRTVSGGQYVVAAAPNLQNQQVLTGLPGVMPNIQYQVIP 177
Chimp     PTSKEQSGSSTNGSNGSESSKNRTVSGGQYVVAAAPNLQNQQVLTGLPGVMPNIQYQVIP 178
Cow       PTSKEQSGSSTNGNNGSESSKNRTVSGGQYVVAATSNLQNQQVLTGLPGVMPNIQYQVIP 177
Dog       PTSKEQSGSSTNGSNGSESSKNRAVSGGQYVVAATSNLQNQQVLTGLPGVMPNIQYQVIP 169
Pig       PTSKEQSGSSTNGNNGGESSKNRTVSGGQYVVAATSNLQNQQVLTGLPGVMPNIQYQVIP 177
Rabbit    PTSKEQSGSNTNGSNGSESSKNRTVSGGQYVVAATPNLQNQQVLTGLPGVMPNIQYQVIP 179
Mouse     PTSKEQSGNSTNGS---ESSKNRTVSGGQYVVAATPNLQNQQVLTGLPGVMPNIQYQVIP 176
Rat       PTSKEQSGNSTNGSNGSESSKNRTVSGGQYVVAATPNLQNQQVLTGLPGVMPNIQYQVIP 178
          ********..***.   ******:**********:.************************

                   .         .         .         .         .
Human     QFQTVDGQQLQFAATGAQVQQDGSGQIQIIPGANQQIITNRGSGGNIIAAMPNLLQQAVP 237
Chimp     QFQTVDGQQLQFAATGAQVQQDGSGQIQIIPGANQQIITNRGSGGNIIAAMPNLLQQAVP 238
Cow       QFQTVDGQQLQFAATGAQVQQDGSGQIQIIPGANQQIITNRGSGGNIIAAMPNLLQQAVP 237
Dog       QFQTVDGQQLQFAATGAQVQQDGSGQIQIIPGANQQIITNRGSGGNIIAAMPNLLQQAVP 229
Pig       QFQTVDGQQLQFAATGAQVQQDGSGQIQIIPGANQQIITNRGSGGNIIAAMPNLLQQAVP 237
Rabbit    QFQTVDGQQLQFATTGAQVQQDGSGQIQIIPGANQQIITNRGSGGNIIAAMPNLLQQAVP 239
Mouse     QFQTVDGQQLQFAATGAQVQQDGSGQIQIIPGANQQIIPNRGSGGNIIAAMPNLLQQAVP 236
Rat       QFQTVDGQQLQFAATGAQVQQDGSGQIQIIPGANQQIITNRGSGGNIIAAMPNLLQQAVP 238
          *************:***********************.**********************

                   .         .         .         .         .
Human     LQGLANNVLSGQTQYVTNVPVALNGNITLLPVNSVSAATLTPSSQAVTISSSGSQESGSQ 297
Chimp     LQGLANNVLSGQTQYVTNVPVALNGNITLLPVNSVSAATLTPSSQAVTISSSGSQESGSQ 298
Cow       LQGLANNVLSGQTQYVTNVPVALNGNITLLPVNSVSAATLTPSSQAVTISSSGSQESGSQ 297
Dog       LQGLTNNVLSGQTQYVTNVPVALNGNITLLPVNSVSAATLTPSSQAVTISSSGSQESGSQ 289
Pig       LQGLANNVLSGQTQYVTNVPVALNGNITLLPVNSVSAATLTPSSQAVTISSSGSQESGSQ 297
Rabbit    LQGLANNVLSGQTQYVTNVPVALNGNITLLPVNSVSAATLTPSSQAVTISSSGSQESGSQ 299
Mouse     LQGLANNVLSGQTQYVTNVPVALNGNITLLPVNSVSAATLTPSSQAGTISSSGSQESSSQ 296
Rat       LQGLANNVLSGQTQYVTNVPVALNGNITLLPVNSVSAATLTPSSQAGTISSSGSQESGSQ 298
          ****:***************************************** *********.***
```

Figure 1. (Continued)

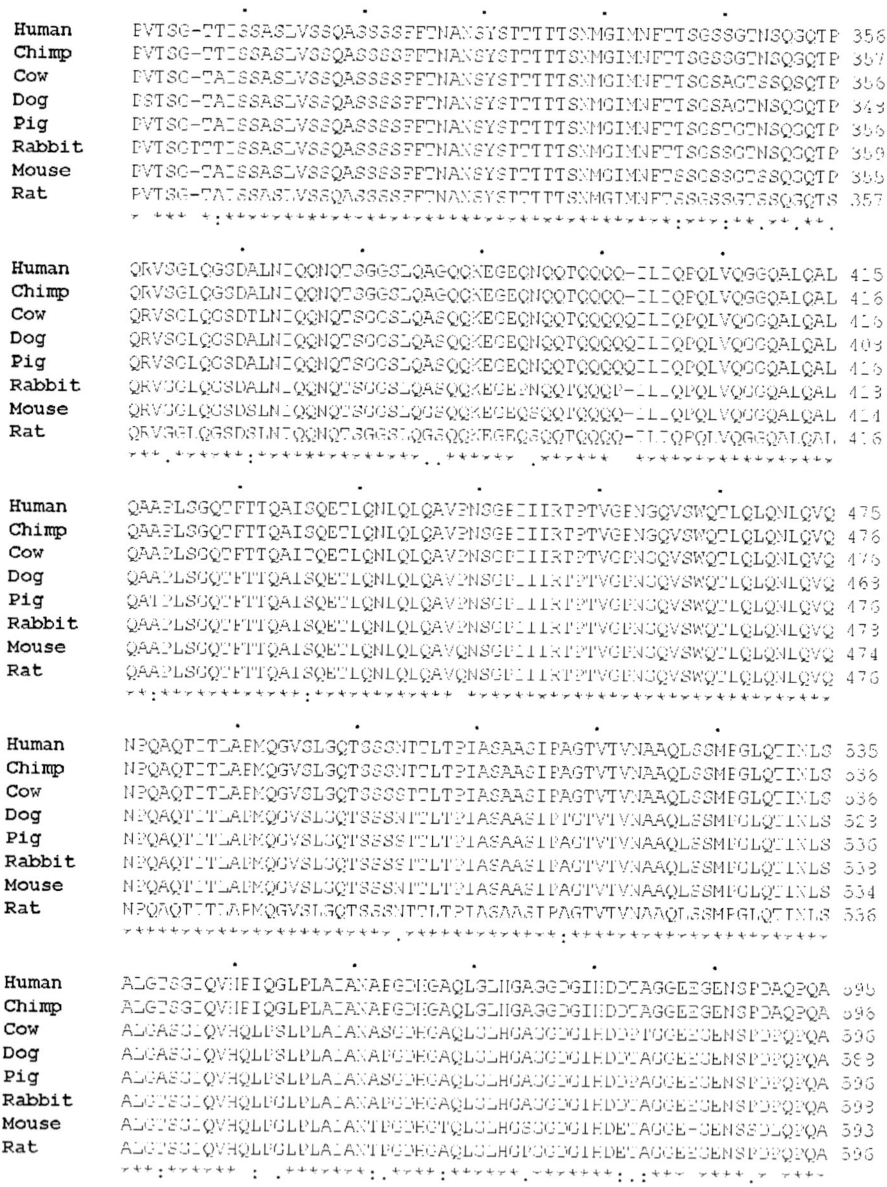


Figure 1. (Continued)

```
                                                                 ZF1
                   .         .         .         .         .
Human    GRRTRREACTCPYCKDSEGRGSGDPGKKKQHICHIQGCGKVYGKTSHLRAHLRWHTGERP 655
Chimp    GRRTRREACTCPYCKDSEGRGSGDPGKKKQHICHIQGCGKVYGKTSHLRAHLRWHTGERP 656
Cow      GRRTRREACTCPYCKDSEGRGSGDPGKKKQHICHMQGCGKVYGKTSHLRAHLRWHTGERP 656
Dog      GRRTRREACTCPYCKDSEGRGSGDPGKKKQHICHIQGCGKVYGKTSHLRAHLRWHTGERP 648
Pig      GRRTRREACTCPYCKDSEGRGSGDPGKKKQHICHMQGCGKVYGKTSHLRAHLRWHTGERP 656
Rabbit   GRRTRREACTCPYCKDSEGRASGDPGKKKQHICHIQGCGKVYGKTSHLRAHLRWHTGERP 658
Mouse    GRRTRREACTCPYCKDSEGRASGDPGKKKQHICHIQGCGKVYGKTSHLRAHLRWHTGERP 653
Rat      GRRTRREACTCPYCKDSEGRGSGDPGKKKQHICHIQGCGKVYGKTSHLRAHLRWHTGERP 656
         ********************.*************:*************************

                    ZF2                           ZF3
                   .         .         .         .         .
Human    FMCTWSYCGKRFTRSDELQRHKRTHTGEKKFACPECPKRFMRSDHLSKHIKTHQNKKGGP 715
Chimp    FMCTWSYCGKRFTRSDELQRHKRTHTGEKKFACPECPKRFMRSDHLSKHIKTHQNKKGGP 716
Cow      FMCTWSFCGKRFTRSDELQRHKRTHTGEKKFACPECPKRFMRSDHLSKHIKTHQNKKGGP 716
Dog      FMCTWSYCGKRFTRSDELQRHKRTHTGEKKFACPECPKRFMRSDHLSKHIKTHQNKKGGP 708
Pig      FMCTWSYCGKRFTRSDELQRHKRTHTGEKKFACPECPKRFMRSDHLSKHIKTHQNKKGGP 716
Rabbit   FMCTWSYCGKRFTRSDELQRHKRTHTGEKKFACPECPKRFMRSDHLSKHIKTHQNKKAGP 718
Mouse    FMCNWSYCGKRFTRSDELQRHKRTHTGEKKFACPECPKRFMRSDHLSKHIKTHQNKKGGP 713
Rat      FMCNWSYCGKRFTRSDELQRHKRTHTGEKKFACPECPKRFMRSDHLSKHIKTHQNKKGGP 716
         ***.**:**************************************************.**

                   .         .         .         .         .
Human    GVALSVGTLPLDSGAGSEGSGTATPSALITTNMVAMEAICPEGIARLANSGINVMQVADL 775
Chimp    GVALSVGTLPLDSGAGSEGSGTATPSALITTNMVAMEAICPEGIARLANSGINVMQVADL 776
Cow      GVALSVGTLPLDSGAGSEGSGTATPSALITTNMVAMEAICPEGIARLANSGINVMQVADL 776
Dog      GVSLSVGTLPLDSGAGSEGSGTATPSALITTNMVAMEAICPEGIARLANSGINVMQVADL 768
Pig      GVALSVGTLPLDSGAGSEGSGTATPSALITTNMVAMEAICPEGIARLANSGINVMQVADL 776
Rabbit   GVALSVGTLPLDSGAGSEGSGTATPSALITTNMVAMEAICPEGIARLANSGINVMQVADL 778
Mouse    GVALSVGTLPLDSGAGSEG--TATPSALITTNMVAMEAICPEGIARLANSGINVMQVTEL 771
Rat      GVALSVGTLPLDSGAGSEGSGTATPSALITTNMVAMEAICPEGIARLANSGINVMQVTEL 776
         **:***************  *************************************::*

                   .
Human    QSINISGNGF 785
Chimp    QSINISGNGF 786
Cow      QSINISGNGF 786
Dog      QSINISGNGF 778
Pig      QSINISGNGF 786
Rabbit   QSINISGNGF 788
Mouse    QSINISGNGF 781
Rat      QSINISGNGF 786
         **********
```

Figure 1. Amino acid sequence conservation of Sp1. Amino acid sequences of Sp1 from human (NP_612482.2), chimpanzee (XP_003313713.1), cow (NP_001071495.1), dog (XP_543633.2), pig (XP_003355454.1), rabbit (XP_002711174.1), mouse (NP_038700.2), and rat (NP_036787.2) were aligned using ClustalW2 software (www.ebi.ac.uk). The level of Aamino acid sequence identity between human Sp1 and that of chimpanzee, mouse, rat, rabbit, cow, and dog is are respectively 99%, 96%, 97%, 96%, 98%, 95%, and 96%. Three $C_2H_2$ ZF domains are highly conserved; their positions are marked and the 2 Cys and 2 His residues are highlighted. Amino acid identity in all species is indicated by *.

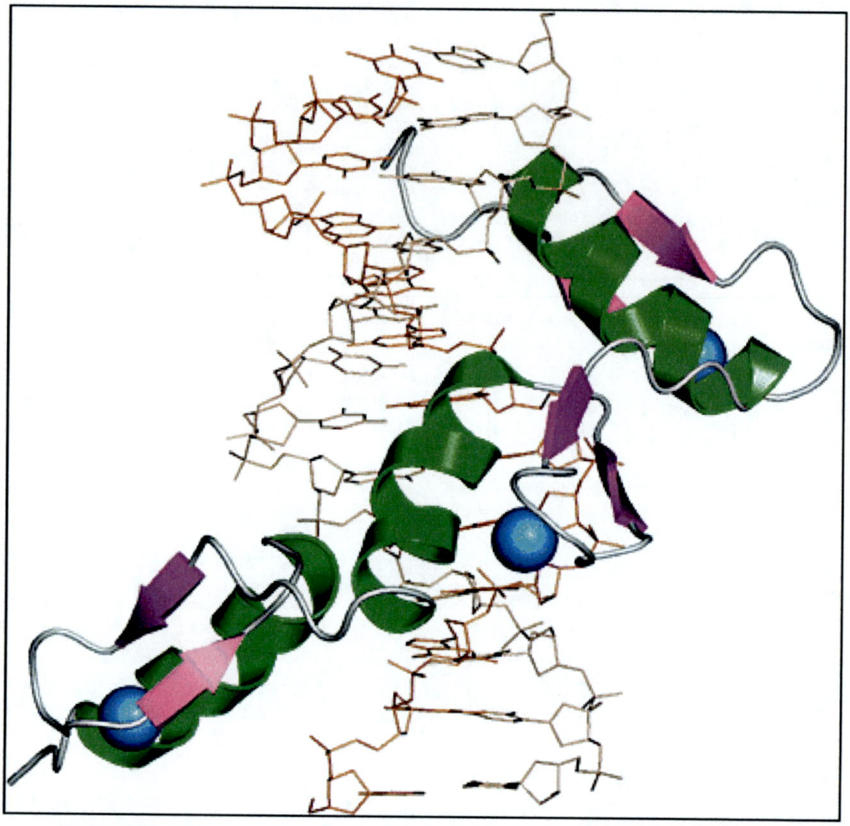

Figure 2. NMR structure of DNA binding domain of Sp1. The proposed ββα zinc finger motifs of Sp1 bind to the DNA. Zn is shown as blue sphere. Figure is from Oka et al. (29) and was used with permission from the journal.

In addition, Sp1 expression has been shown in COS-7 cells [28], which is a kidney fibroblast-like cell line derived from African green monkey, as well as in a pig proximal tubule cell line known as LLC-$PK_1$ [18]. Similarly, rat kidney epithelial cell line, NRK-52E, was shown to express Sp1 [58,54], and we have shown its mRNA and protein expressions in primary cultures of mouse kidney cells [48]. Overall, Sp1 is present in the renal tissues and kidney cell lines from different species.

Amino acid sequence alignment shows that human Sp1 is highly homologous to Sp1 from non-human primates, rodents and other mammals (Figure 1). Therefore, the commonly used kidney cell lines HK-2, HEK-293, COS-7, NRK-52E, and LLC-$PK_1$ can be used to evaluate Sp1 mediated promoter activity by reporter assay method or to isolate nuclear Sp1 to examine its binding to an oligonucleotide probe by Electrophoretic Mobility Shift Assay (EMSA). It is expected that these cells provide suitable models to examine the regulation of Sp1-driven gene expression *in vitro*.

# Kidney Glucose Transporters

Normally, glucose is freely filtered out of the blood at the glomerulus of the kidney nephron but then it is almost completely reabsorbed in the proximal tubule region of the nephron by sodium-dependent glucose transporters SLC5A1 and SLC5A2, commonly known as SGLT1 and SGLT2, respectively [56]. SGLT2 is responsible for reabsorption of most of filtered glucose [51] while SGLT1 takes up the residual glucose from the glomerular filtrate [52]. Uptake of glucose by SGLTs is coupled to the transport of sodium into the cell which in turn is driven by the sodium electrochemical gradient that exists across the apical membrane [55]. The relative low concentration of sodium inside the cell is maintained by $Na^+/K^+$ ATPase, which pumps out excess sodium [56]. Absorbed glucose then is transported out of the cell into blood from the basolateral side via facilitative glucose transporters SLC2A1 and SLC2A2, also known as GLUT1 and GLUT2, respectively.

# Regulation of *SGLTs* by Sp1

## *SGLT1*

The gene encoding SGLT1 is located on human chromosome 22 (Table 1). Analysis of 1000 base pairs (bp) chromosomal DNA sequence upstream of human *SGLT1* (h*SGLT1*) gene shows 4 predicted Sp1 binding sites within the region 300 bp immediately upstream of the first Methionine codon of thise gene (Figure 3). Luciferase reporter assay analysis with several chromosomal clones ranging from ~5 kilo bases (kb) to less than 300 bp showed that all of these fragments exhibited promoter activity in kidney LLC-$PK_1$ cells; however, had stronger activity in human intestinal cell line Caco-2 [24].

SGLT1 is a major glucose transporter in the small intestine [55], and characterization of h*SGLT1* promoter has so far been carried out in Caco-2 cell line [24]. The minimal promoter of h*SGLT1* gene is localized in a ~250 bp region 5' of the gene; within this region, and the two 3′ GC-boxes, named GC Box 1 and 2, are both essential for its basal activity *in vitro* [24]. The positions of these sites are shown in Figure 3A and their sequences are highlighted in Figure 3B.

Promoter region of rabbit *SGLT1* (r*SGLT1*) has also been isolated (20). Unlike h*SGLT1* promoter, GC-1 site appears not play a role in r*SGLT1* promoter activity examined in Caco-2 cells. However, similar to h*SGLT1* promoter, the down-stream GC-box or GC-2 site, is essential for basal activity of r*SGLT1* promoter.

**Table 1. *SGLTs* chromosomal locations and analyzed promoter regions**

| **SGLT1** | **Chromosome** | **Strand** | **Reference Assembly No.** | **Contig No.** | **Start (5′)** | **End (3′)** |
|---|---|---|---|---|---|---|
| Human | 22 | + | NC_000022.10 | NT_011520.12 | 11828838 | 11829837 |
| Mouse | 5 | + | NC_000071.5 | NT_039302.7 | 379738 | 380737 |
| Rat | 14 | - | NC_005113.2 | NW_047430.2 | 400734 | 399735 |
| Pig | 14 | + | NC_010456.3 | NW_003536226.1 | 1165210 | 1166209 |
| **SGLT2** | | | | | | |
| Human | 16 | + | NC_000016.9 | NT_010393.16 | 31433458 | 31434457 |
| Mouse | 7 | + | NC_000073.5 | NT_039433.7 | 46008378 | 46009377 |
| Rat | 1 | + | NC_005100.2 | NW_047562.2 | 30131762 | 30132761 |

We isolated a 4.6 kb mouse chromosomal clone which contains both the coding and 1.3 kb upstream sequences of *SGLT1* (m*SGLT1*) gene [48]. Comparison between human minimal promoter and nucleotide sequences upstream of m*SGLT1* shows that there is 72% identity between these sequences (Figure 3B and ref. 48), and that the locations and core sequences of GC-boxes as well as their flanking sequences are highly conserved (Figure 3A and 3B). In addition to rabbit and mouse, the GC-1 and GC-2 sites are highly conserved in chromosomal regions upstream of *SGLT1* genes from pig and rat (Figure 3A and 3B).

EMSA experiments on DNA binding reactions with a recombinant human Sp1 (rhSp1) and oligonucleotide probes with sequences of mouse GC-1 and GC-2 sites show that rhSp1 can bind to both sequences [48]. Nuclear proteins isolated from primary cultures of mouse kidney cells (PMKC) form similar complexes as rhSp1 with the same probes suggesting that nuclear Sp1 is bound. Evidence for presence of Sp1 in these complexes is shown by appearance of supershifts in migration of protein/probe complexes when Sp1 antibody is added to the binding reactions [48].

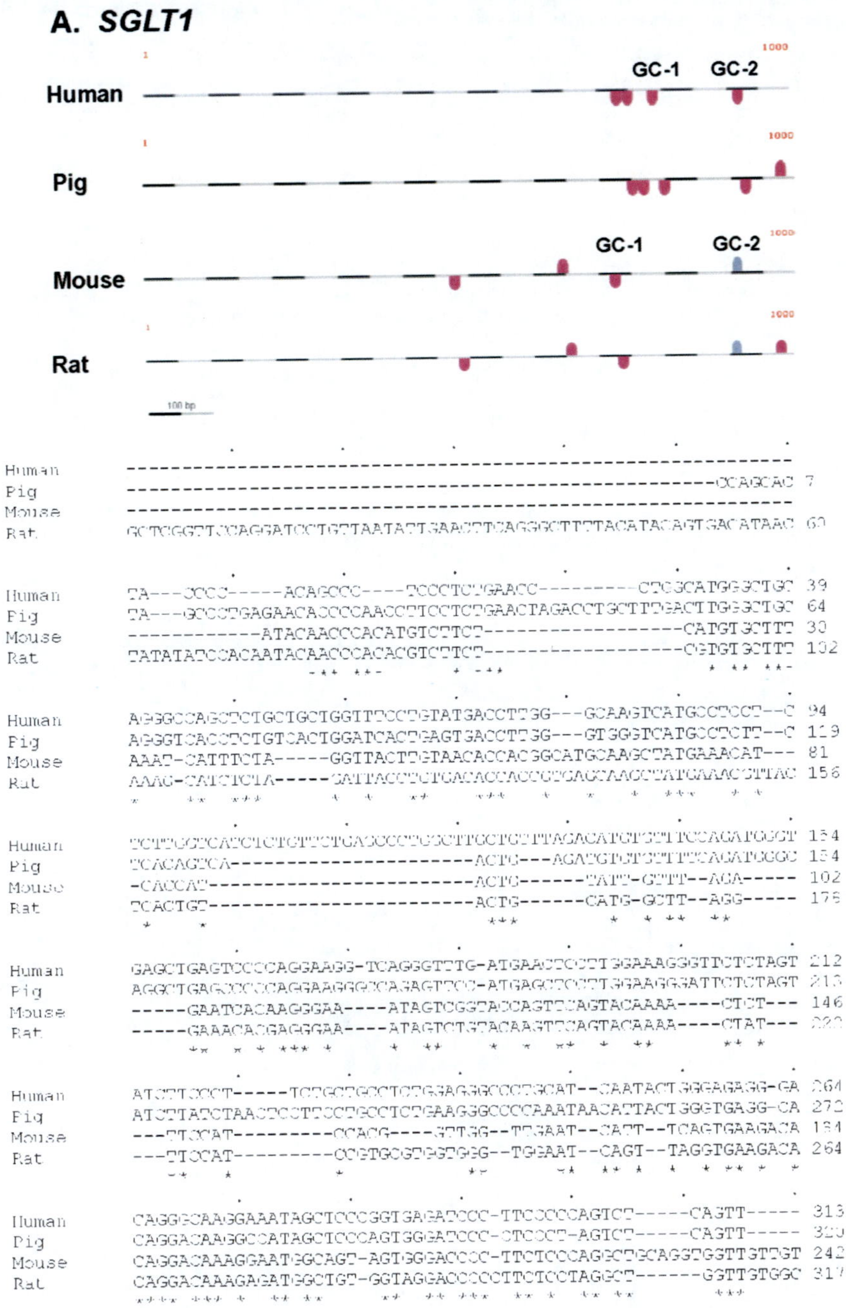


Figure 3. (Continued)

```
Human   TCTT------CCTCTTTA--------------CAGTGG--------------------- 331
Pig     TCTT------CTTCCCTA--------------CAGTGG--------------------- 338
Mouse   TCTTTCTGTGCTCCCTCAGTGGGTAGCTCCCGCAGTGGGTAGCTCCCGCAATTGCTCTTA 302
Rat     TCT----------CCTTA--------------CAGTGGT-------------------- 332
        ***        *   *              ******

Human   ------GGGTAACAGTGGACCTGGGGGCTGCAGGCTGTGGAGGGGGAGTCAGGAATGTGG 385
Pig     ------GGGTTACAGTGCTCCTGGGAGCTGCAGGCTATAGCGAGGGAGGCAGAAAGGTGA 392
Mouse   CGGGGGGAACTGCACTGTTTCTAGGGTCTA-ATGCTGTAGAGG----TTCGGGAA--TGG 355
Rat     ----GGGAATCGCGCAGTTCCTAGGGATTGCATGCT-----------CTGGGGAA----- 372
              *     *   *   ** **   *  * ***                * **

Human   GTGGGTGGGTGGCTGTCCTTGTGGCCAAGAGATTAGGATTGTGCTGGTTTTTTTCCAGTG 445
Pig     ACGG-TGAGTGGCCATCCTTGTGGCTAAGAGAATGCCATTGTGCTGAATGTTTTCCAGCA 451
Mouse   GTG--TCTGTG---ATGCTGGTGGTCAAGAGATT-----TGTACTGACTTTCTCCCAGCA 405
Rat     --G--TCAGTA---ACG---GTGGCTAAGGGATT-----TGTATTGCGTTTTCCCCAGCC 417
          *  *  **          ****  *** ** *      ***  **  * *   ****

Human   ACACCCAGCAACTCTCTGTCCTCTCAAGAATGTTCTCCAGGGAGAACAAGA-CTTGACTG 504
Pig     ACATCCAACAGCTTATGGTCCTCTACAGAATGTTCTCCATGGAGACCAAGATCTTGGCTG 511
Mouse   CCGTCCAGCAGTCTC--ATCCTCCTGACCAT-TTCTCCA-GGAGACCAAGAACTTGGCTG 461
Rat     CCATCCAGCAATCTC--ATCCTCCTGAGAAT-TTCTCCA-GGAGACCAAGAACTTGGCTG 473
         *  *** **         *****   *  ** ******* ***** ***** **** ***

Human   ATTCTTCAGAAGAGTGCTGCACCCATTTTACAAGCGGACAAGCCAAGTCCTTAGAAGGCA 564
Pig     ATTCCT-GCAACAATCCTGCACCCATTTTACACACGGAGAAGCCAGGTCCACGGAGAACA 570
Mouse   AGTTCTCTGAGAAGTCCTGCACTCCTCTTTC--ACAGAGACGGCA-GTCCACAAAGATGC 518
Rat     AGTTCTCTGCAAAGTCCTGCACTCGTTTTTC-CACAGAGAAATCA-GTCCACAGAGATGC 531
        * *  *       * * ****** * * ** *   * ** *   ** ****    *

Human   TGGGACTAGTCCAAGGATGGGTG--GCAGGTGATCTGAAGCCTCTAGGCTCCTAGCCCCA 622
Pig     CAAGACTGGTCCAAGGGCATATG--GCAGGTGATCTGAAGCTTCTAGGCCCCTAGCACCA 628
Mouse   CAGGTCTGCTTCAAAAGCACCTGCCGCAGATGGGCCCCAGCCTATAGATGCTTGGCTCCA 578
Rat     CAGGTCTGCTTCAAAAGCACCTGCCACAGATAAGCCTCAGCCTATAGACCCTTGGCTCCA 591
           * **  * ***        **    *** *   *   *** * ***   * * ** ***

Human   GGTGGC-CTGTTCCCTCAATGGAGCCACTCCAGCTCTGGC-------------AAGGGCA 668
Pig     G-CAGA-CTGACTCCTCAATGGAGTCACTCCAGCTCTGACCTGGGTGTGGCCCAAGAGCA 686
Mouse   GGTGCCTCTGGTCCCTCTGTGTAGCAACTCCTTTTCTGCCTCACGTGGGACTCCGTGGCA 638
Rat     GGTGCCTCCG-TCCCTCTGTGTAGCTACTCCTTTTCTGCTTCACGTGGGACTCCGTGGCA 650
        *      * *    ****  ** **  *****   ****                   ***

Human   GGTGTGGTGGCGC-----TGC-TGGTGCCTGCTGTGGGAGTACAGTGGGCCTTTGTGCCA 722
Pig     GGCCAGGTGAGGC-----TGCCTGATGCCTGCTGCATGAGCACAGCGGGCCTTTGTGCCA 741
Mouse   GACCAGGCGTGGCAACTGTTCGAGGTGCCCACTGCAGGAACACAGAGGACTTTTGTGCCA 698
Rat     AGCCAGGCGTGGCAACTGTGCGAGGTGCCCACTGCGGGAGCACAGAGGACTTTTGTGCCA 710
             ** *  **       * *  * ****  ***    **   **** ** * *********

                                   GC-1

Human   GGCGGGCACCACCCCCTGGCTGGCCCCTCCCCATTCGCAGGACAGCTCTTACCTGCCGGG 782
Pig     GGCTGGCACCACCCCCTGGCTAGCCCCTCCCTGATCCAGAGACAGCCAGTACCTACCCGC 801
Mouse   GGCAGTCACCACCCCTTAGCAGGCCCCTCCCTGCTCACAGAACAGACTTTACCTGCCGGC 758
Rat     GGCGGTCACCACCCCTTAGCCGGCCCCTCCCTGCTCACCGAACAGACTTTACCTGCCGGC 770
        *** * ********* * **  *********   **       ****     ***** ** *

Human   ---------------CCGCCGCCCCA--GCCAACAGCTCAGCCGGGTGCTCCTTCCTGGG 825
Pig     ---------------CCTCCACCTCCTTGCCAATAGCTCTACCAGCTCCTCCTC-----G 841
Mouse   CAGGCGCCCAGGCGCCCGCAGACTCT---CTTG-AGCTTTGCTCGGT-CTTCTTAAGGCA 813
Rat     -------CCGGGCGCCCGCAGTCTCT---CTTGCAGCTTTGCTGA------CTTAGAGCA 814
                       ** *   * *    *    ****    *          **

Human   CTCCACGCCCGG-AGCTGCTTCCTGACGGTGCAGCC--GCAAGGCATCGCAGGGGCCCCG 882
Pig     TTCCGCACTCAG-ACCTGCTCCCTGACGGTGCAG--------GGC----TAGGGCCCCTG 888
Mouse   CTGCACACCTAGGAGCTGCTTCCCGACGGTGCGGTGCAGCGCAGCAGCACAGGGGCCCAG 873
Rat     CTGCACACCTAGGAGCTGCTTCCCGACGGTGCGGTGCAGCGCAGCAGCCCAGGGGCCCAG 874
          * * * *    * * ***** ** ******** *            **    **** *** *
```

Figure 3. (Continued)

**GC-2**

```
                   .         .         .         .         .         .
Human     CGCTACTGCCCTGCTCCCTCAAAGTCCCAGGTCCCCTCCCCTGGTGCTGATCATTAACCA 942
Pig       CATGACAGCCCTGCTGCCACACT-TCACGGGTCCCCTCCCCTGAGGCTGATCATTAACCC 947
Mouse     CGTTCCTGCCGCGCAGCCCCACTGTCGCGGGTCCCCTCCCTCGGGGCTGATCATTAACTC 933
Rat       CGTTCCTGCCGCGCAGCCCCTCCGTCGCGGTTCCCCTCCCTCGGGGCTGATCATTAACTC 934
          *    * ***  **  ** *    ** * * *********  *  *************

                   .         .         .         .         .         .
Human     GGAGGCCGTATAAGGAGCTAGCGGCCCTGGCGAGA---------GGGAAGGACGCAACGC 993
Pig       AGAGGCCACAAAAGAAGCAAGTGGCTTGGGTG-----------------GGACGCATTGT 990
Mouse     AAAAGCAGTATAAGGAGTGAGTGGCCCCGGTGAGCCTGTGAGCCGAGAGGGACTCGTCGC 993
Rat       AAAAGCAGTATAAGGAGTGAGTGGCCCCGGTGAGCCGGTGAGCCG-GAGGGACTCGTCGC 993
            * **   * *** **  ** ***   ** *                 **** *   *

                   .
Human     TGC---CACC 1000
Pig       TGCAGTCGCC 1000
Mouse     CAC---CGCC 1000
Rat       CAC---CGCC 1000
            *   * **
```

## C. *SGLT2*

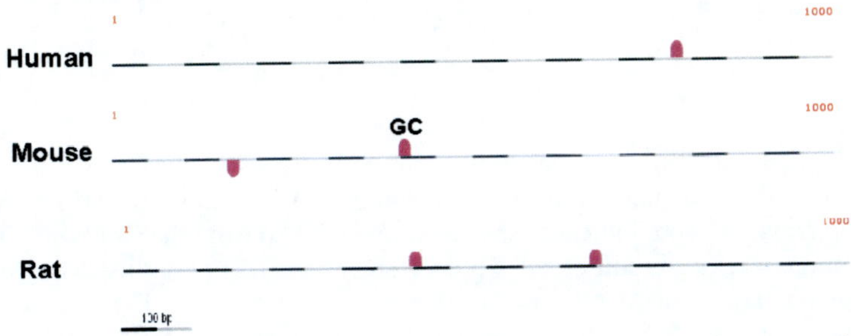


```
                   .         .         .         .         .         .
Mouse     GGAAAATGAGATGTGGGCACAAAGTCCCACTCCTAAGTAAGAAACTACTTGCA-GTTGAC 59
Rat       -----ACAGGAGATGG----AAAG----ATTTTTAAG-----AGCCAG----A-GTTGGA 37
Human     ------TATTAACCAGCC-----TCTCCCTTCTTGAAT------CCAGTCCTGTGCTGAG 43
                    *    *              *  * *        * *       * **

                   .         .         .         .         .         .
Mouse     AGCTA---CTAGGAGAGAGAAATGGGT-----------TTTTT----TCAATGGAG-AGA 100
Rat       GGTTGT--CTCCAAGAGAGCA--GAGT-----------TTTCCAGACACAACAGAGCGGA 82
Human     AAATGAGACTCCAGGGATGAATCAGATCAGTCCTGCTCTTTTTAATTTTTATTTATTTAT 103
             *    **    *   * *     *           ***         *   *

                   .         .         .         .         .         .
Mouse     CACTGGGTGTATCAACT-GCACCCCAAGGCAGGCCACAC--GCTCAGGA----AGAGTTG 153
Rat       CGC---GTATCTGAACTCACAGACCATGGCAGCCCATACAAGACCCGCA----TGAGTT- 134
Human     TTTTTAAGAGATGAGGTCTCACTCC---GCTGCCCAGGCTGGTTTCAAACTCCTGAGCT- 159
                     * *  *  **  **   ** * ***  *  *      *     *** *

                   .         .         .         .         .         .
Mouse     GCCAACACAAAACACACTCCGTGTT-TGGTTTTCT-GTTTGCTTGGGTT-TGTTTTTGTG 210
Rat       --CAA-ACCAAATACAATCCCAGCA-TGGAAA----ATAAGAAGGGGCA-CAAAGTTCCA 185
Human     --------CAAGTGAGCCTCCTGCTGCAGCCTCCCAAAGTGCTAGGATTACAGGCATGAG 211
                   **        *  *     *           *   **          *

                   .         .         .         .         .         .
Mouse     CTTTTATTC------TTCCTTCCTTT----------------CTTTCTTTGTTTCTTTGT 248
Rat       CTCCTAACCAAGAAACGACAGCCACTGGGAGAGAGAAATGGGCTTTCTTCAGTGGATAGA 245
Human     CCCCCACAC--------CCAGCCAGT----------------CCTACTCT--TGAGGAGA 245
          *    *  *         *  **  *                * * **    *     *
```

Figure 3. (Continued)

```
Mouse    TTCTCTCTCTCTTTCTCTCTCTTTTTCTCTCCCTCCTCCTCCTCCTCTTCTTTCTTCTTT 308
Rat      TGCTGGGTATATCAACTGCACCCCAAGTCAGGCCACACGCTCAGGCCAACACAAAACATA 305
Human    CACTAGTT-TGTTGGGAAGAC---AGACCTGGACCCAGGCAAGGGCCA---CACAGTGTG 298
           **   * * *         *       *      *          *           *

Mouse    TTCTTCTTCTTCT-TCTTCTTCTTCTTCTTCTTCTTCTTCTTCTTCTTCTTCTTCTTCTT 367
Rat      CTC--CGTGTTCTGTTTTCTTTTGCTTTGGTCTGTTTTTGTGCTTTTCATTTTTTTTTCTT 363
Human    CTC--------------AGAACTACATCAGAAATGTCAGAGAAGTCT--------TTCTT 336
          **               * * *        *        * *        *****

Mouse    CTTCTTCTTCTTCTTCTTCTTCTTCTGAT-CAGAGAGGGGAG-GGGATCTGGG-AAAAGT 424
Rat      TTGAGAGAGAAAAAGGATATGAAGTTGGT-TTGAGAGGGGAG-GGGGTCTGGGGAGAAGT 421
Human    GAAGAGGCGACACAGGCTG-------GGTACTGAAAGCTGAGTAGGAGTTGGC---CATT 386
                            *       * *  ** **  ***  **  ***     * *

Mouse    TGGGGGAGAGGAAAGAATT-TGCCCAAAATATA----TTGCATAAAAATTTTTAAAAAT 479
Rat      TGAGGGAGAGGAAAGAATT-TGCCCAAAATATA----TTGCATGAAAACATTTAAAAAC 476
Human    TGGGAGACAGGGAAGTGCTCTAAACAGAAGACACAGTTTGTGCAAGGATTTGGTGGGGA- 445
         ** * ** *** ***    * *   ** ** * *   ***   *  *  *  *    *

Mouse    CCTTCCCCCAGGGGAATCCAGGGACTAGCCCCTTGA-GGGACAGATGCTGGAGAGA 1000
Rat      CCTTCCCC-AGGAGATTCCAGGCACTAGCCCCCTTGA-GGGGCAGATGCTGGAGAGA 1000
Human    CCTTCCCCTGGGGGAATCCAGGGGCTGGTTCCTGGATGGGGCAGATCCTGGGGAGA 1000
         ********  ** ** ******  ** *  *** ** *** ***** **** ****
```

Figure 3. Sp1 binding sites upstream of *SGLT1* and *SGLT2* genes. One kb chromosomal DNA sequences upstream of the first ATG codon of *SGLT1* (A,B) and *SGLT2* (C,D) genes from indicated species were obtained from the GenBank (Table 1). Transcription factor binding sites were predicted by MatInspector software (www.genomatix.de) and nucleotide sequences were aligned using ClustalW2 (www.ebi.ac.uk). A,C- Schematic presentations of predicted Sp1 (pink) and Zn07 (blue) matrix families. A box represents gene's coding region. GC, GC-1, GC-2 indicate sites to which Sp1 binding has been experimentally shown (21,24,48). B,D- Nucleotide sequences of the predicted Sp1 and Zn07 factor sites are highlighted and tandem sites are boxed. Matrix for Sp1 is highlighted in dark gray and that of Zn07 is in light gray. Sequences are numbered in 5′ to 3′ direction relative to the first ATG codon.

Our Chromatin Immunoprecipitation (ChIP) assay on cultured mouse kidney cells show that Sp1 occupies chromosomal region encompassing GC-1 site *in vivo* [21].

Despite a high level of sequence conservation in GC-2 regions from different species (Figure 3B), transcription factor binding analysis on mouse and rat sequences assigns matrix family zinc finger 7 (ZF07), instead of Sp1, to this region and predicts that zinc finger protein 263 [10] binds to it (Figure 3A). However, our Chromatin Immunoprecipitation (ChIP) assay on cultured mouse kidney cells show that Sp1 occupies chromosomal regions encompassing GC-1 and supershift assay shows that nuclear Sp1 from PMKC binds to GC-2 sites upstream of m*SGLT1* promoter *in vitrovo* (4821). Taken together, these results suggest that GC-2 region is a conserved binding site for Sp1.

### *SGLT2*

*SGLT2* gene is located on human chromosome 16 (Table 1). There is one predicted Sp1 binding site approximately 230 bp upstream of h*SGLT2* (Figure 3C); however, it is not yet known whether Sp1 regulates h*SGLT2* gene expression. This GC box is not conserved in chromosomal sequences upstream of mouse or rat *SGLT2* genes (Figure 3D). However, each mouse and rat sequence has two predicted GC boxes but only one of these sites is conserved and was called GC site [21]. DNA binding assay and EMSA analysis using m*SGLT2*-GC oligonucleotide probe showed that nuclear Sp1 from cultured mouse kidney cells bind to this probe [21]. Consistently, *in vivo* binding of Sp1 protein to the chromosomal region encompassing this conserved GC-box was shown in PMKC [21]. These results suggest that Sp1 may play a role in regulation of human *SGLT2* gene, and this hypothesis needs further investigation.

## Regulation of *GLUT*s by Sp1

Role of Sp1 in regulation of kidney expression of *GLUT1* (*SLC2A1*) or *GLUT2* (*SLC2A2*) promoter gene has not so far been investigated. However, Sp1 may play roles in regulation of *GLUT1* gene expression in skeletal muscle [39], liver [13], trophoblast [30], and adipose (19) cells. Sp1 binding site upstream of the rat *GLUT2* gene has been reported [1] but it is not yet known whether expression of this gene is regulated by Sp1.

### *GLUT1*

Transcription factor binding site analysis of 1 kb chromosomal DNA sequences upstream of human *GLUT1* gene (Table 2) predicts that there are 9 Sp1 matricxes, 7 of which are clustered within a region 300-400 bp upstream of the first ATG codon (Figure 4A). The two most 3′ sites are not conserved in mouse and rat (Figure 4A). However, similar to human, there are respectively

6 and 8 tandem Sp1 sites upstream of mouse and rat *GLUT1* genes, respectively. The locations and nucleotide sequences of human, mouse, and rat Sp1 cluster sites are highly conserved (Figure 4B). Based on the above, we predict that binding of Sp1 to these tandem sites play roles in regulation of *GLUT2* gene expression, and this hypothesis requires further investigation.

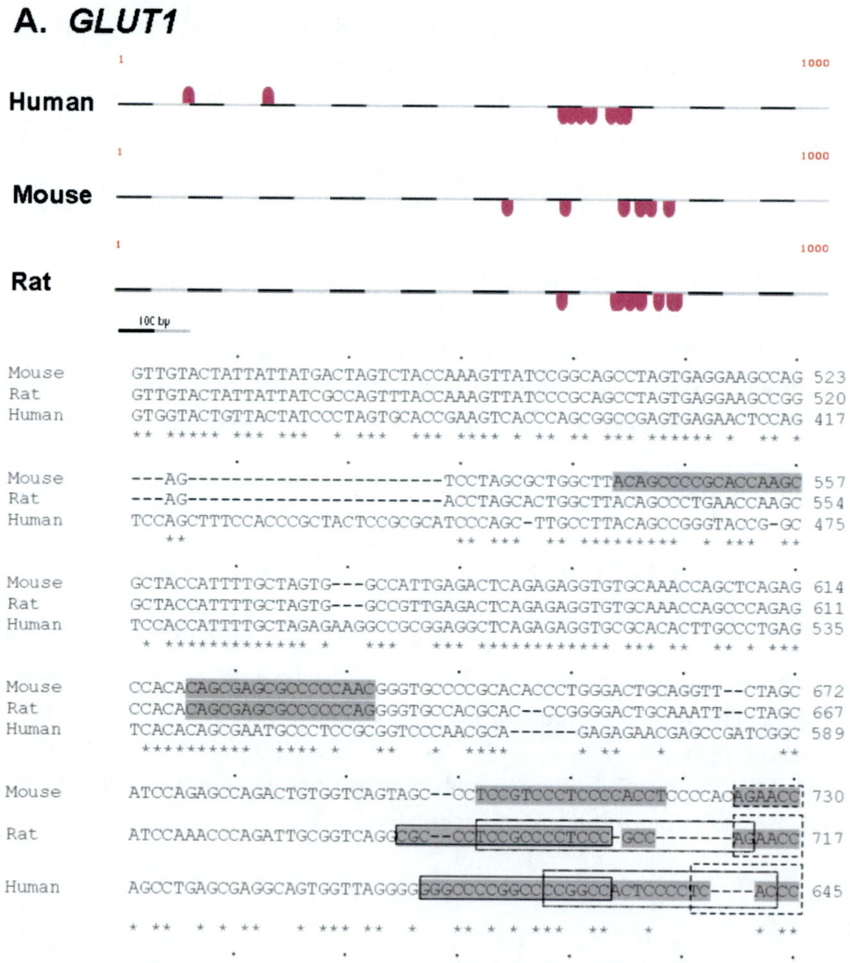


Figure 4. (Continued)

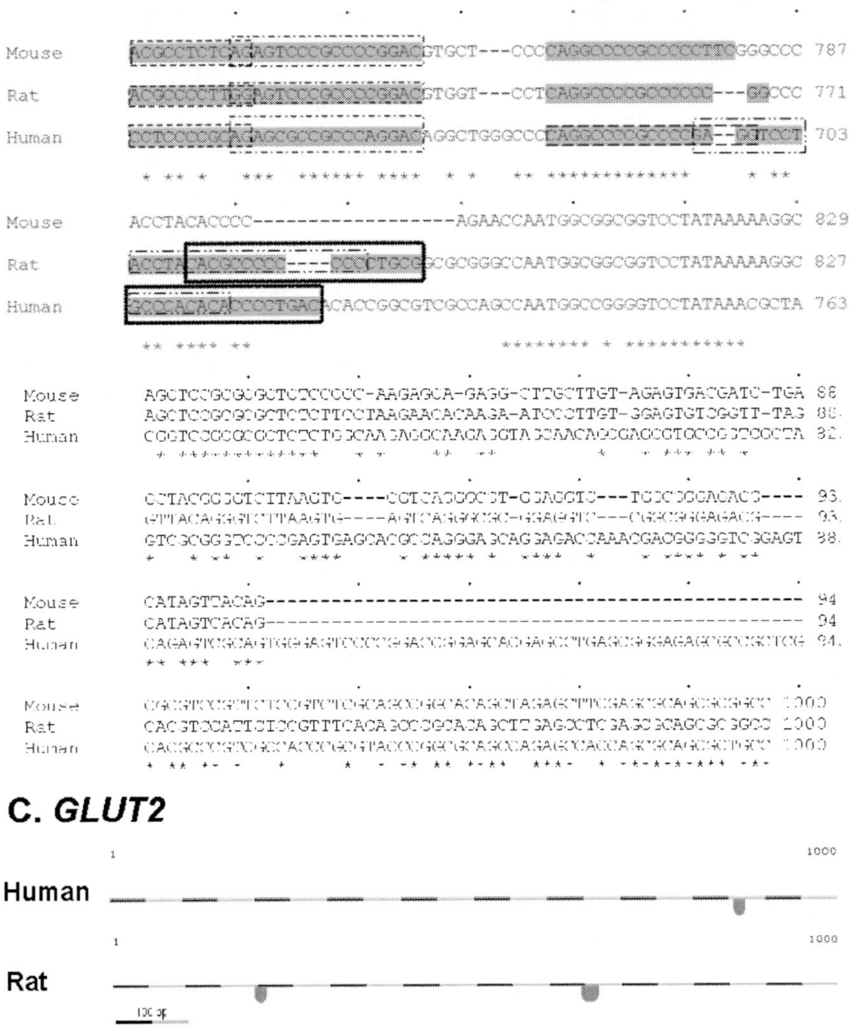


Figure 4. (Continued)

```
                 .         .         .         .         .         .
Mouse   ------------------------------------------------------------
Rat     CCTCTCACATTTCCTCTGCTGCAAACTGGCCCCGTAAGGACTCTGTGGGTGCCAGTAACC 60
Human   ---TACATTTTTGTTCCGCTGGGAACTGGTTCCATTTGAGCCTCAGTGCAATCTGTGGCC 57

                 .         .         .         .         .         .
Mouse   ---GCTGGCAGTCCCAGG-GAAGCTACCCAAACTTTGGTATCTCCCCCAGTATTTTTGTA 56
Rat     ACAGCCGCCAGTCACAGT-GAAGCTACACTAACTTTGGTATCTCCCTCATTATTTTTGGA 119
Human   CCAGTTGCCACACACACTTAAAGCT-CAATAAATTTGATATCTCCTGAACCCCCACAGGG 116
           *  * **  * **    ***** *   ** **** *******   *        *

                 .         .         .         .         .         .
Mouse   TC-AGATGCCACAGGGGATTCCTAGGAAGAAACAGGGTCAAGACACGTAAGCAGAATTCA 115
Rat     TC-AGATGCCAGAGGGGATTCCTAGAAAGAAACAAGATGAAGACATGTGAGTAGAATTTA 178
Human   TTTGGATATCATAAGGGATACTTAGG----CCCAAGATCAAGCCATATGAGCAGG-TTTA 171
        *   ***  ** * ***** * ***       ** * * *** **  * ** **  ** *

                 .         .         .         .         .         .
Mouse   CTGGAAAGCCAAGGCTATAGCAAGCACACTACGGCTCCCAACCAGGCTCTCTATGTGTAG 175
Rat     TAGGAAACCCAAGGCTATAACAAACACACTCCCTCTCCCAATCAGGGTCTCTATGTGTAG 238
Human   TTTGAGAAC----------------------------------------TCTACGT---- 187
           ** * *                                        **** **

                 .         .         .         .         .         .
Mouse   CCAAGGCTATCCTACAACTCTCATGTGATCCAGTGAGTCTCCAATGCA------------ 223
Rat     CCAAGGCTGTCCTGCAACTCCCATGCCATCCAATGAGCCTCCAACTCAGCCCTCCTGTCC 298
Human   -TAAAGCC--------------------TCCAGCAAGT---------------------- 204
          ** **                     ****   **

                 .         .         .         .         .         .
Mouse   CAGCCTCCCCAGTTAGAGCATTATAGATGTGCGCCACCGCATCTGGCTCCGCACTCTCAT 283
Rat     CAGCCTCCCCAGTGTGAGCATTACAGATGTGT--CACCCCATCCGGCTGCGTG-TCTCAT 355
Human   ---------------------------------------------GCT-----------T 208
                                                     ***           *

                 .         .         .         .         .         .
Mouse   CTTGAGATACACTGACTTAATAATAACAGTAGAAAGCACATTGCCACTATCTACTGAGAA 343
Rat     CCTAAAATACACTGAGTTAATAACAACAG---AAAGCATGTGGCCACCATCTACCAAGAA 412
Human   CTTGACATATA-TGAAGCAATAATAATAA-------CA----ACCGCTATTTGCTAAGTA 256
        * * * *** * ***   ***** ** *        **     ** * ** * *  ** *

                 .         .         .         .         .         .
Mouse   CTTTTTCTG----------TGCCAAACATTTAT-CACGTTATTTCCCTTAATCCCCTCAA 392
Rat     CTTTTTTTGT--------ATGCCAAACATTTAT-CGTGCTATTTCGTTTAATCCCCTCAA 463
Human   CTTATTATATGTTGAGATATGCCAAACACTTTCACATATTATCTCATTTAATCCACTCAA 316
        *** ** *           ********* **   *    *** **  ******* *****

                 .         .         .         .         .         .
Mouse   CAAAACAAAATGAGTGTGCTATCTTCCCTATTTTATCAGGCCAAAAATCAAACAGTAGAA 452
Rat     CAAAACAGAATGCCTGTGCTATTATCTCTATTTGGTCAAGTCAGAAACCAAACAATAGAA 523
Human   CAAAACAAAATG--TGTGTTATT---TCTATTTTATCAGG--------CAAA-AATAAAA 362
        ******* ****  **** ***     ******  *** *        **** * ** **

                 .         .         .         .         .         .
Mouse   ACCTAAGACACAGAAAAGTCACAGGGTTTGACTCGCTTAAGTTCAT---CCTCACTAAAA 509
Rat     AACTAAGACACAGAAAAGTCACAG------------------------------------ 547
Human   AACTAAT-CAGAGAAAAGTCACAGTGAGCCACTTGCTTAAGGTTATACTCCCCAGTAAAA 421
        * ****  ** *************

                 .         .         .         .         .         .
Mouse   TGCTGTGATTCCAACCCCAGTGCAGATCCCT------CCACACCTCACACAGATA----A 559
Rat     --------------------TGCAGACCCCT------CCGGCCCTCATTCAGATAGAAAA 581
Human   TGTTGAGATTCAAA------CCCAAGTCCCTAACAATCCAAAGCTCATCCAGGTAACCAT 475
                              **   ****      **    ****  *** **

                 .         .         .         .         .         .
Mouse   CCGATGCTGCCACACTCTGTGGCCACAGAGC-------------------CCACAGAACT 600
Rat     CCAATGCCTCCACAGTCTGTGGCTACAGAGC-------------------CCACAGCACT 622
Human   TCAAGTCAACAATGGTCTCCGGCCACACAGCTCAGCTCTAAACAGAAACACCACAGCACT 535
         * *  *  * *   ***  *** *** ***                   ****** ***
```

Figure 4. (Continued)

```
Mouse     AATTTTCTGCAGAACAAACTGCTCTGTGCCCTCC-------CAACTTTGTCTTCCCGACA 653
Rat       AGTTCTCTGCAGCGCAAACTGCTCTGCACCCTCCCCCCTCCCCACTTTGGCTTCCCCACA 682
Human     AATTATCTGTGGAGCAGAGTGCAGTGTGCCTCCCATGCT--CCACAGCA-CAGCACGGCA 592
          * ** ****  *  ** * ***  **  **  **        * **     *  * *  **

Mouse     CCACTGCATGCCTCGGCTTCCACAAAAGACCCCACAGGCTGCCCCTCTCAGCTGTCCCTG 713
Rat       ACACAGCATGGCTCAGCTTCTCTAAGATACCCTACAGGTTGCTCCTCTCACCTGTCCCCA 742
Human     GCTCAGCATATCTCATCTTCTACAAAAGACTCCAAAGATTTCTCTTTTCACCAGCTCCCA 652
             * * ****  ***  ****   ** * ** * * **  * * * * *** * *   **

Mouse     TCCCATTTCTGCCACACTTATCAGGC-TGAAAATGGGTCTGTCTCTGGTTTGTAACTTAT 772
Rat       TCCCATTTCTGCCACATTCATCAGCCCTTAAAATGGTTCTGCCTCTGGCTTGTAACTTAT 802
Human     ----ATTACTGCCACATAACACATGCCTTGAAATGTGATTGCCTCTGGTTTGTAACTTAT 708
              *** ********      **  * *  *****    ** ****** ***********

Mouse     ACTTATGAGACCTGCTACTGTGCTC------------------------AAGCCACAAGT 808
Rat       ------GAGACCTCCTATTGTTCTC------------------------AAGCCACAGGT 832
Human     GCCTAAGGGACCTGCTCCCATTTTCTTTCCTAGTGGAACAAAGGTATTGAAGCCACAGGT 768
               * ***** **    *  **                         ******** **

Mouse     CATTGGGGTAAAGGGTGTATTGATTGGATTACCATCAATACTCAGCTTCTGTTAAAAAGG 868
Rat       CA----------AGGTATATTGATTGAATTACCATCAATACGCAGCCTTGGTTAAGAAGG 882
Human     TGCTGAGGCAAAGCACTTATTGATTAGATTCCCATCAATATTCAGCTGCCGCTGAGAAGA 828
                        ********  *** *********  ****     * * * ***

Mouse     TCAGAACT-A-----------------------------CCTCTCTTTGCTCCTCCTCCT 898
Rat       TCAGAATTCA-----------------------------CCTCCGCTTGCTCCTCCTCCT 913
Human     TTAGACTTGGACTCTCAGGTCTGGGTAGCCCAACTCCTCCCTCTCCTTGCTCCTCCTCCT 888
          * ***  *                               ****   *************

Mouse     CCTACAATGTATAACCAGGT------AGAGT---GAGCACTCTGGCTGGTCAGCTATTCA 949
Rat       ---ACAATGTATAACCAGGC------AGAG-----------CTGA--GGACAGCTACTCA 951
Human     ---GCAATGCATAACTAGGCCTAGGCAGAGCTGCGAATAAACAGGCAGG--AGCTAGTCA 943
              ***** ***** ***         ****            * *   **  ***** ***

Mouse     T---------CCACATTCAGTACAGGACCTGGA-TTAAGAGGACAATTCCACACACACAC 999
Rat       T---------CCACATTCGGAACAGGACCTGGA-TTGAGACAACAACTCCGCACGCA-AC 1000
Human     GGTGCATGTGCCACACTCA-CACAAGACCTGGAATTGACAGGAC---TCCCAACTAGTAC 999
                    ***** **   *** ******** ** * *  **   ***  **     **

Mouse     A 1000
Rat       -
Human     A 1000
```

Figure 4. Sp1 binding sites upstream of *GLUT1* and *GLUT2* genes. One kb chromosomal DNA sequences upstream of the first ATG codon of *GLUT1* (A,B) and *GLUT2* (C,D) genes from indicated species were obtained from the GenBank (Table 2). Prediction of transcription factor binding sites and sequence alignments were performed as described in Figure 3. A box represents gene's coding region. A,C- Schematic presentations of Sp1 matrix. Sp1 binding site was not predicted in mouse. B,D- Nucleotide sequences of the predicted Sp1 factor sites are highlighted. Sequences are numbered in 5′ to 3′ direction relative to the first ATG codon.

### *GLUT2*

Sequence analysis predicts that there is a Sp1 binding site in close proximity to the human *GLUT2* (h*GLUT2*) gene (Figure 4C and 4D); however, this site is not conserved in mouse or rat (Figure 4D). Analysis of current GenBank chromosomal DNA sequences indicates that there is no Sp1 binding site upstream of mouse *GLUT2* gene (Table 2 and Figure 4D).

**Table 2. SGLTs chromosomal locations and promoter regions**

| SGLT1 | Chromosome | Strand | Reference Assembly No. | Contig No. | Start (5′) | End (3′) |
|---|---|---|---|---|---|---|
| Human | 22 | + | NC_000022.10 | NT_011520.12 | 11828838 | 11829837 |
| Mouse | 5 | + | NC_000071.5 | NT_039302.7 | 379738 | 380737 |
| Rat | 14 | - | NC_005113.2 | NW_047430.2 | 400734 | 399735 |
| Pig | 14 | + | NC_010456.3 | NW_003536226.1 | 1165210 | 1166209 |
| **SGLT2** | | | | | | |
| Human | 16 | + | NC_000016.9 | NT_010393.16 | 31433458 | 31434457 |
| Mouse | 7 | + | NC_000073.5 | NT_039433.7 | 46008378 | 46009377 |
| Rat | 1 | + | NC_005100.2 | NW_047562.2 | 30131762 | 30132761 |

# Role of Sp1 in Metal-Induced Glucosuria

## Cadmium-Induced Nephrotoxicity

Cadmium (Cd) is a heavy metal and a major environmental pollutant [16]. Cd has no role in biological functions and is considered a nonessential metal [49]. Ingestion of contaminated food [42], inhalation of polluted air [16], and cigarette smoking [36] are major non-occupational routes of human exposure to Cd. The majority of circulating Cd ends up in the liver where it induces metallothionein (MT), a low-molecular weight metal binding protein [38]. Cd-MT complex is formed and released into the blood stream and is filtered through the kidney glomerulus [38]. Kidney is the major site of Cd accumulation in the body and the renal cortex contains 20 to 40 fold higher levels of Cd than the liver [41]. In human, cadmium levels ranging from 2 to 70 μg/g wet kidney weight [2,40] or 130 μg/g dry cortical tissue weight have been reported [32]. Cadmium concentrations of r 1μg/g wet kidney weight have been associated with renal dysfunction [41].

Cadmium-induced nephrotoxicity can result in acquired Fanconi-like syndrome, which is manifested by abnormal urinary excretion levels of essential nutrients including glucose [43]. Loss of glucose in the urine, or the symptom of glucosuria, as a result of exposure to Cd is consistent with preferential accumulation of this toxic metal in the proximal tubule region of the kidney (4), the major site of glucose reabsorption.

## Kidney Cell Culture Model

Primary mouse kidney cell culture (PMKC) has been used in our laboratory as an *in vitro* model to study the mechanism of Cd-induced glucosuria [47]. Similar to its effects on human kidney, exposure of cultured kidney cells to μM concentrations of Cd results in inhibition of sodium-dependent uptake of glucose. This adverse effect of Cd on glucose uptake is dose-dependent and is manifested in response to exposure of cells to concentrations as low as 3 μM [7]. Cadmium exhibits the same inhibitory effects on sodium-dependent glucose transport in LLC-$PK_1$ cells [7].

## mRNA Expression of SGLTs in Cd-Exposed Kidney Cells

Cultured mouse kidney cells express mRNA of *SGLT1* and *SGLT2* [47]; the mRNA expression of *SGLT1* has also been shown in LLC-$PK_1$ (9) but it is not yet known whether *SGLT2* is expressed in these cells. Inhibition of sodium-dependent glucose transport by Cd is suggestive of down-regulation of SGLT in Cd-exposed cells. Consistent with this hypothesis, we have shown that exposure of PMKC to increasing concentrations of Cd results in dose-dependent decreases in mRNA levels of *SGLT1* [47] in cells in which *SGLT1* mRNA stability is not affected [48]. In these cells, Cadmium exposure causes similar decreases in mRNA levels of *SGLT2* in cultured mouse kidney cells [47].

## Effect of Cd on Sp1 Binding to GC Sites in SGLT Promoters

While mRNA and protein levels of Sp1 in Cd-exposed PMKC are similar to their levels in un-exposed cells [48], Cd treatment causes significant decreases in *in vitro* binding of nuclear Sp1 to m*SGLT1*-GC-1 and GC-2 probes [48]. Compared to un-treated cells, binding of nuclear Sp1 to m*SGLT2*-GC-box probe is also reduced in response to Cd [21]. Consistently, Cd exposure results in decreases in *in vivo* bindings of Sp1 to the chromosomal regions encompassing m*SGLT1*-GC-1 and m*SGLT2*-GC-box siteequences of m*SGLT1* and m*SGLT2*, respectively [21].

## Effect of Cadmium on Zinc Finger DNA Binding Domain of Sp1

The zinc (Zn) ion in the zinc finger (ZF) domain of Sp1 can be replaced by Cd; however, the consequence of such metal replacement on Sp1 binding activity has been inconsistent. The DNA binding activity of a bacterially expressed recombinant peptide encompassing three ZF domains of Sp1 was the same when this peptide was allowed to fold in presence of either Zn or Cd [22]. In another study, reconstitution of rhSp1 apoprotein with Cd resulted in an active DNA binding protein [50]. On the other hand, results from other studies including that of ours suggest otherwise and are discussed below.

Presence of Cd in reactions with Zn-containing rhSp1 (Zn-rhSp1) and a GC consensus probe results in down-regulation of Sp1 binding [34]. Our results show that *in vitro* exposure of Zn-rhSp1 to increasing concentration of Cd causes dose-dependent decreases in subsequent binding of rhSp1 to the m*SGLT1*-GC1 and m*SGLT2*-GC probes [21]. However, the level of Zn-rhSp1 binding to these probes remains un-altered if Cd is added after the transcription factor and DNA are allowed to interact. Similarly, the activity of rhSp1 bound to m*SGLT1*-GC-1 probe, but not free rhSp1, is protected against Zn chelation by apo-metallotheionin [33]. These results strongly support that down-regulation of Sp1 activity by Cd is in part due to direct interaction of Cd with the zinc finger DNA binding domain of Sp1.

## Blockade of the Inhibitory Effects of Cd by Zn

Zinc supplementation during Cd exposure has protective effects on proximal tubule cells *in vitro* and *in vivo*. Cadmium-induced apoptosis and necrosis in a rabbit PT cell line could be blocked by co-exposure to Zn [3]. Also, development of Fanconi-like symptoms of increased urinary excretions of sodium, potassium and other nutrients in rats injected with Cd could be prevented by co-treatment of animals with Zn [15]. Similarly, addition of equimolar concentrations of Zn during Cd exposure significantly protects against Cd-induced decreases in mRNA levels of *SGLT1* and *SGLT2* in primary cultures of mouse kidney cells [21].

EMSA shows that rhSp1 co-treated with Cd and Zn has stronger binding to m*SGLT1*-GC1 and m*SGLT2*-GC sequences than rhSp1 that is exposed to Cd alone [21]. However, Zn supplementation following Cd treatment is not protective. Therefore, the availability of Zn to the pool of un-bound Sp1 may protect against the inhibitory effect of Cd on Sp1 binding to the *SGLT*s GC sites.

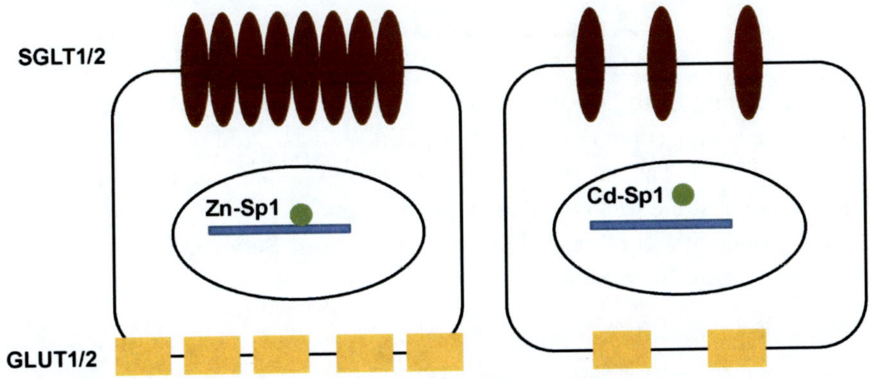


Figure 5. Predicted mechanism of cadmium-induced glucosuria. Reabsorption of filtered glucose in the nephron involves SGLT1 and SGLT2, which are localized on the apical side of the proximal tubule cells. Reabsorbed glucose is then transported back to the blood from the basolateral side via GLUT1 and GLUT2. We predict that human exposure to Cd results in down-regulation of binding of the ubiquitous transcription factor Sp1 to the GC-box sequences which can result in decreased expressions of *SGLT*s and *GLUT*s genes ultimately leading to decreases in the number of these glucose transporters in the kidney proximal tubule causing symptom of glucosuria.

# Conclusion

The exact mechanism of down-regulation of glucose transport in the kidneys of cadmium-exposed humans is not known. However, results suggest that it may be due to the adverse effects of this toxic metal on activity of zinc finger transcription factor Sp1, which is known to regulate *SGLT1* gene and there are predicted sites for its binding in the chromosomal regions upstream of *SGLT2*, *GLUT1*, and *GLUT2* genes (Figures 3 and 4). Our working model, which is presented in Figure 5, predicts that cadmium exposure ultimately leads to decreases in expressions of SGLT1 and 2 on the apical and GLUT1 and 2 on the basolateral side of the kidney proximal tubule cells leading to inhibition of glucose reabsorption and the development of glucosuria.

## Acknowledgments

The National Institutes of Health grants R01ES004026 and P30ES004184 (to D.H. Petering) supported our referenced studies. Additional funds were provided (to N.M. Tabatabai) by the Department of Medicine and the Kidney Disease Center of the Medical College of Wisconsin.

A.B. Blodgett's current affiliation is U.S. Forest Service, Forest Products Laboratory; Madison, Wisconsin.

## References

[1] Ahn YH, Kim JW, Han GS, Lee BG and Kim YS. Cloning and characterization of rat pancreatic beta-cell/liver type glucose transporter gene: a unique exon/intron organization. *Arch Biochem Biophys* 323: 387-396, 1995.

[2] Baker JR, Satarug S, Reilly PE, Edwards RJ, Ariyoshi N, Kamataki T, Moore MR and Williams DJ. Relationships between non-occupational cadmium exposure and expression of nine cytochrome P450 forms in human liver and kidney cortex samples. *Biochem Pharmacol* 62: 713-721, 2001.

[3] Barbier O, Dauby A, Jacquillet G, Tauc M, Poujeol P and Cougnon M. Zinc and cadmium interactions in a renal cell line derived from rabbit proximal tubule. *Nephron Physiol* 99: 74-84, 2005.

[4] Bernard A. Renal dysfunction induced by cadmium: biomarkers of critical effects. *Biometals* 17: 519-523, 2004.

[5] Black AR, Black JD and Azizkhan-Clifford J. Sp1 and kruppel-like factor family of transcription factors in cell growth regulation and cancer. *J Cell Physiol* 188: 143-160, 2001.

[6] Blumenthal S, Lewand D, Krezoski SK and Petering DH. Comparative effects of Cd2+ and Cd-metallothionein on cultured kidney tubule cells. *Toxicol Appl Pharmacol* 136: 220-228, 1996.

[7] Blumenthal SS, Lewand DL, Buday MA, Kleinman JG, Krezoski SK and Petering DH. Cadmium inhibits glucose uptake in primary cultures of mouse cortical tubule cells. *Am J Physiol* 258: F1625-F1633, 1990.

[8] Chu S and Ferro TJ. Sp1: regulation of gene expression by phosphorylation. *Gene* 348: 1-11, 2005.

[9] Clancey CJ and Lever JE. Differential regulation of three glucose transporter genes in a renal epithelial cell line. *J Cell Physiol* 185: 244-252, 2000.

[10] Frietze S, Lan X, Jin VX and Farnham PJ. Genomic targets of the KRAB and SCAN domain-containing zinc finger protein 263. *J Biol Chem* 285: 1393-1403, 2010.

[11] Graham FL, Smiley J, Russell WC and Nairn R. Characteristics of a human cell line transformed by DNA from human adenovirus type 5. *J Gen Virol* 36: 59-74, 1977.

[12] Hakvoort TB, Moerland PD, Frijters R, Sokolovic A, Labruyere WT, Vermeulen JL, Ver Loren vT, Breit TM, Wittink FR, van Kampen AH, Verhoeven AJ, Lamers WH and Sokolovic M. Interorgan coordination of the murine adaptive response to fasting. *J Biol Chem* 286: 16332-16343, 2011.

[13] Hwang DY and Ismail-Beigi F. Control of Glut1 promoter activity under basal conditions and in response to hyperosmolarity: role of Sp1. *Am J Physiol Cell Physiol* 290: C337-C344, 2006.

[14] Jackson SP and Tjian R. O-glycosylation of eukaryotic transcription factors: implications for mechanisms of transcriptional regulation. *Cell* 55: 125-133, 1988.

[15] Jacquillet G, Barbier O, Cougnon M, Tauc M, Namorado MC, Martin D, Reyes JL and Poujeol P. Zinc protects renal function during cadmium intoxication in the rat. *Am J Physiol Renal Physiol* 290: F127-F137, 2006.

[16] Jarup L. Hazards of heavy metal contamination. *Br Med Bull* 68: 167-182, 2003.

[17] Kaczynski J, Cook T and Urrutia R. Sp1- and Kruppel-like transcription factors. *Genome Biol* 4: 206, 2003.

[18] Kajiwara M, Terada T, Asaka J, Ogasawara K, Katsura T, Ogawa O, Fukatsu A, Doi T and Inui K. Critical roles of Sp1 in gene expression of

human and rat H+/organic cation antiporter MATE1. *Am J Physiol Renal Physiol* 293: F1564-F1570, 2007.

[19] Kao YS and Fong JC. Endothelin-1 induces glut1 transcription through enhanced interaction between Sp1 and NF-kappaB transcription factors. *Cell Signal* 20: 771-778, 2008.

[20] Kekuda R, Saha P and Sundaram U. Role of Sp1 and HNF1 transcription factors in SGLT1 regulation during chronic intestinal inflammation. *Am J Physiol Gastrointest Liver Physiol* 294: G1354-G1361, 2008.

[21] Kothinti RK, Blodgett AB, Petering DH and Tabatabai NM. Cadmium down-regulation of kidney Sp1 binding to mouse SGLT1 and SGLT2 gene promoters: possible reaction of cadmium with the zinc finger domain of Sp1. *Toxicol Appl Pharmacol* 244: 254-262, 2010.

[22] Kuwahara J and Coleman JE. Role of the zinc(II) ions in the structure of the three-finger DNA binding domain of the Sp1 transcription factor. *Biochemistry* 29: 8627-8631, 1990.

[23] Li L, He S, Sun JM and Davie JR. Gene regulation by Sp1 and Sp3. *Biochem Cell Biol* 82: 460-471, 2004.

[24] Martin MG, Wang J, Solorzano-Vargas RS, Lam JT, Turk E and Wright EM. Regulation of the human Na(+)-glucose cotransporter gene, SGLT1, by HNF-1 and Sp1. *Am J Physiol Gastrointest Liver Physiol* 278: G591-G603, 2000.

[25] Michael DR, Phillips AO, Krupa A, Martin J, Redman JE, Altaher A, Neville RD, Webber J, Kim MY and Bowen T. The human hyaluronan synthase 2 (HAS2) gene and its natural antisense RNA exhibit coordinated expression in the renal proximal tubular epithelial cell. *J Biol Chem* 286: 19523-19532, 2011.

[26] Narayan VA, Kriwacki RW and Caradonna JP. Structures of zinc finger domains from transcription factor Sp1. Insights into sequence-specific protein-DNA recognition. *J Biol Chem* 272: 7801-7809, 1997.

[27] Niu Y, Mo D, Qin L, Wang C, Li A, Zhao X, Wang X, Xiao S, Wang Q, Xie Y, He Z, Cong P and Chen Y. Lipopolysaccharide-induced miR-1224 negatively regulates tumour necrosis factor-alpha gene expression by modulating Sp1. *Immunology* 133: 8-20, 2011.

[28] Nores R, Blanchon L, Lopez-Diaz F, Bocco JL, Patrito LC, Sapin V and Panzetta-Dutari GM. Transcriptional control of the human pregnancy-specific glycoprotein 5 gene is dependent on two GT-boxes recognized by the ubiquitous specificity protein 1 (Sp1) transcription factor. *Placenta* 25: 9-19, 2004.

[29] Oka S, Shiraishi Y, Yoshida T, Ohkubo T, Sugiura Y and Kobayashi Y. NMR structure of transcription factor Sp1 DNA binding domain. *Biochemistry* 43: 16027-16035, 2004.

[30] Okamoto Y, Sakata M, Yamamoto T, Nishio Y, Adachi K, Ogura K, Yamaguchi M, Takeda T, Tasaka K and Murata Y. Involvement of nuclear transcription factor Sp1 in regulating glucose transporter-1 gene expression during rat trophoblast differentiation. *Biochem Biophys Res Commun* 288: 940-948, 2001.

[31] Philipsen S and Suske G. A tale of three fingers: the family of mammalian Sp/XKLF transcription factors. *Nucleic Acids Res* 27: 2991-3000, 1999.

[32] Pulido P, Kagi JH and Vallee BL. Isolation and some properties of human metallothionein. *Biochemistry* 5: 1768-1777, 1966.

[33] Rana U, Kothinti R, Meeusen J, Tabatabai NM, Krezoski S and Petering DH. Zinc binding ligands and cellular zinc trafficking: apo-metallothionein, glutathione, TPEN, proteomic zinc, and Zn-Sp1. *J Inorg Biochem* 102: 489-499, 2008.

[34] Razmiafshari M and Zawia NH. Utilization of a synthetic peptide as a tool to study the interaction of heavy metals with the zinc finger domain of proteins critical for gene expression in the developing brain. *Toxicol Appl Pharmacol* 166: 1-12, 2000.

[35] Razmiafshari M, Kao J, d'Avignon A and Zawia NH. NMR identification of heavy metal-binding sites in a synthetic zinc finger peptide: toxicological implications for the interactions of xenobiotic metals with zinc finger proteins. *Toxicol Appl Pharmacol* 172: 1-10, 2001.

[36] Richter PA, Bishop EE, Wang J and Swahn MH. Tobacco smoke exposure and levels of urinary metals in the U.S. youth and adult population: the National Health and Nutrition Examination Survey (NHANES) 1999-2004. *Int J Environ Res Public Health* 6: 1930-1946, 2009.

[37] Ryan MJ, Johnson G, Kirk J, Fuerstenberg SM, Zager RA and Torok-Storb B. HK-2: an immortalized proximal tubule epithelial cell line from normal adult human kidney. *Kidney Int* 45: 48-57, 1994.

[38] Sabolic I, Breljak D, Skarica M and Herak-Kramberger CM. Role of metallothionein in cadmium traffic and toxicity in kidneys and other mammalian organs. *Biometals* 23: 897-926, 2010.

[39] Sanchez-Feutrie M, Santalucia T, Fandos C, Noe V, Vinals F, Brand NJ, Ciudad CJ, Palacin M and Zorzano A. A novel muscle DNA-binding activity in the GLUT1 promoter. *Cell Mol Life Sci* 61: 709-720, 2004.

[40] Satarug S, Haswell-Elkins MR and Moore MR. Safe levels of cadmium intake to prevent renal toxicity in human subjects. *Br J Nutr* 84: 791-802, 2000.

[41] Satarug S, Nishijo M, Ujjin P, Vanavanitkun Y, Baker JR and Moore MR. Evidence for concurrent effects of exposure to environmental cadmium and lead on hepatic CYP2A6 phenotype and renal function biomarkers in nonsmokers. *Environ Health Perspect* 112: 1512-1518, 2004.

[42] Satarug S, Garrett SH, Sens MA and Sens DA. Cadmium, environmental exposure, and health outcomes. *Cien Saude Colet* 16: 2587-2602, 2011.

[43] Soderland P, Lovekar S, Weiner DE, Brooks DR and Kaufman JS. Chronic kidney disease associated with environmental toxins and exposures. *Adv Chronic Kidney Dis* 17: 254-264, 2010.

[44] Spengler ML, Guo LW and Brattain MG. Phosphorylation mediates Sp1 coupled activities of proteolytic processing, desumoylation and degradation. *Cell Cycle* 7: 623-630, 2008.

[45] Suske G. The Sp-family of transcription factors. *Gene* 238: 291-300, 1999.

[46] Swiergosz-Kowalewska R. Cadmium distribution and toxicity in tissues of small rodents. *Microsc Res Tech* 55: 208-222, 2001.

[47] Tabatabai NM, Blumenthal SS, Lewand DL and Petering DH. Differential regulation of mouse kidney sodium-dependent transporters mRNA by cadmium. *Toxicol Appl Pharmacol* 177: 163-173, 2001.

[48] Tabatabai NM, Blumenthal SS and Petering DH. Adverse effect of cadmium on binding of transcription factor Sp1 to the GC-rich regions of the mouse sodium-glucose cotransporter 1, SGLT1, promoter. *Toxicology* 207: 369-382, 2005.

[49] Thevenod F. Nephrotoxicity and the proximal tubule. Insights from cadmium. *Nephron Physiol* 93: 87-93, 2003.

[50] Thiesen HJ and Bach C. Transition metals modulate DNA-protein interactions of SP1 zinc finger domains with its cognate target site. *Biochem Biophys Res Commun* 176: 551-557, 1991.

[51] Vallon V, Platt KA, Cunard R, Schroth J, Whaley J, Thomson SC, Koepsell H and Rieg T. SGLT2 mediates glucose reabsorption in the early proximal tubule. *J Am Soc Nephrol* 22: 104-112, 2011.

[52] Vallon V. Molecular determinants of renal glucose reabsorption. Focus on "Glucose transport by human renal Na+/D-glucose cotransporters SGLT1 and SGLT2". *Am J Physiol Cell Physiol* 300: C6-C8, 2011.

[53] Waby JS, Chirakkal H, Yu C, Griffiths GJ, Benson RS, Bingle CD and Corfe BM. Sp1 acetylation is associated with loss of DNA binding at promoters associated with cell cycle arrest and cell death in a colon cell line. *Mol Cancer* 9: 275, 2010.

[54] Wang Y, Rangan GK, Goodwin B, Tay YC and Harris DC. Lipopolysaccharide-induced MCP-1 gene expression in rat tubular epithelial cells is nuclear factor-kappaB dependent. *Kidney Int* 57: 2011-2022, 2000.

[55] Wright EM, Martin MG and Turk E. Intestinal absorption in health and disease--sugars. *Best Pract Res Clin Gastroenterol* 17: 943-956, 2003.

[56] Wright EM, Loo DD and Hirayama BA. Biology of human sodium glucose transporters. *Physiol Rev* 91: 733-794, 2011.

[57] Wu N, Siow YL and O K. Ischemia/reperfusion reduces transcription factor Sp1-mediated cystathionine beta-synthase expression in the kidney. *J Biol Chem* 285: 18225-18233, 2010.

[58] Zhou Q, Lin S, Tang R, Veeraragoo P, Peng W and Wu R. Role of Fosinopril and Valsartan on Klotho Gene Expression Induced by Angiotensin II in Rat Renal Tubular Epithelial Cells. *Kidney Blood Press Res* 33: 186-192, 2010.

# Index

## A

## D



## E

## F



## G

## K



## L



## M

## N



## O

## P

## Q



## R



## S

## T

## U



## V



## W



## X



## Y